grains

Also in this series

Rice: Seventy-eight Recipes for
Arborio, Basmati, Brown, Jasmine,
White, and Wild Rice

Beans: Seventy-nine Recipes for
Beans, Lentils, Peas, Peanuts,
and Other Legumes

grains

Seventy-six

Healthy

Recipes for

Barley, Corn,

Rye, Wheat,

and Other

Grains

Joanne Lamb Hayes
and Bonnie Tandy Leblang

Harmony Books/New York

Published by Harmony Books, a division of Crown Publishers, Inc.,
201 East 50th Street, New York, New York 10022.
Member of the Crown Publishing Group.
Random House, Inc.   New York, Toronto, London, Sydney, Auckland
HARMONY and colophon are trademarks of Crown Publishers, Inc.
Manufactured in the United States of America

Library of Congress Cataloging-in-Publication Data
Hayes, Joanne Lamb.
Grains: seventy-six healthy recipes for barley, corn, rye,
wheat, and other grains / Joanne Lamb Hayes and Bonnie Tandy
Leblang. — 1st ed.
Includes index.
1. Cookery (Cereals) 2. Grain. I. Leblang, Bonnie Tandy
II. Title.
TX808.H39 1995   94-19399
CIP
ISBN 0-517-59204-5
10 9 8 7 6 5 4 3 2 1
First Edition

contents

acknowledgments

Thanks—

*to Valerie Kuscenko for her enthusiasm and guidance in the
production of* Grains;

*to Wendy Schuman for her energy and patience in directing the
promotion of our work;*

*and to Bryan, Eric, Heather, and Claire for exploring the infinite
variety of grains with us.*

introduction

Grains and the products made from them nourished the world's population long before anyone discovered how to cultivate them or recorded their importance. As grains developed from wild grass to cultivated crop, each area of the world embraced the one that thrived in local conditions, so that even today, the culinary traditions of an area are defined by the grain product that is closest to the hearts of its people.

In times of plenty, grains are fed to animals, and meat becomes the center of the meal. In times of adversity, grains save the day and support the human population, nestling into the soul of the culture and providing comfort when all else fails. After several generations of meat-centered meals, grains are once again being "discovered" as the excitement in the center of the plate.

Grains are versatile, inexpensive, and conveniently quick cooking (or slow cooking but delicious reheated), and current nutrition research points to the importance of increasing the amount of grains in the diet.

Grains are clearly a healthy addition to the diet, and this collection of recipes will help you prepare them in ways that will inspire you to eat them over and over again.

As a logical extension of our work with *Rice* (Harmony, 1991) and *Beans* (Harmony, 1994), we now take a look at using the true grains—barley, corn, kamut, millet, oats, rye, spelt, teff, triticale, and wheat—and several nongrains that are used in the same ways as grains—amaranth, buckwheat, and quinoa. We keep our concern for a healthy diet in mind while presenting the traditional, heartwarming, and enticing flavors that have made grains such an important part of the world's sustenance.

grain legends and lore

Early civilizations often flourished around golden fields of wild grasses, which offered highly nutritious seed heads. At least as far back as 9000 B.C., efforts were made to cultivate this useful resource by saving some of the seeds, returning them to the ground, and tending to their needs until they could be harvested and stored for use during the winter. As the various grasses and nongrasses (amaranth, buckwheat, and quinoa) became invaluable staples to local populations, they were endowed with mystical properties and heroic histories.

Native to Mexico, *amaranth* was revered by the Aztecs. It played an important part in their religious ceremonies, and Montezuma's soldiers believed this high-energy substance made them invincible. When the

Spanish conquistador Hernán Cortés discovered the importance of amaranth, he ordered all fields burned and declared the possession of even one seed grounds for punishment. Fortunately for us, his success lasted less than five hundred years, and amaranth has lived up to its name, which means "immortal."

It is thought that *barley* was the first grain to be successfully cultivated. A native of Ethiopia, it appears in the early records of all the Mediterranean cultures. Mild mannered but hardy and reliable, barley was Europe's grain of choice until the sixteenth century.

Although it probably originated in central Asia, *buckwheat,* which is really a fruit and not a grain, achieved popularity in northern Europe, where it grows abundantly. Many of the traditional dishes of Russia and Scandinavia are based on this flavorful product.

Corn was a cultivated Western Hemisphere staple for centuries before European explorers discovered it being used in many of the same ways we enjoy it today. The basis of all original American cultures, corn is the subject of many myths. Its native name, *maize,* means "our life," which indicates its importance.

A prehistoric grain of Asia, *millet* nourished the earliest humans and is mentioned in the Bible as the "gruel of endurance."

Originally a grain of northern Europe and north-central Asia, *oats* has been on American breakfast tables since the seventeenth century. It is still usually found at breakfast, but times are changing.

Just as amaranth sustained the Aztec culture, *quinoa* was the high-energy staple of the Incas until the Spanish arrived and destroyed it. Not as successfully banned as amaranth, quinoa has continued to enjoy popularity in the Andes and is now gaining fans all over the world.

Rye has been the mainstay of northern countries for nearly two thousand years because it grows well in climates that cannot support wheat. Rye added its distinctive flavor to the traditional breads of northern Europe and provided sustenance to New England colonists who were not able to produce wheat successfully.

In the late nineteenth century, the first attempt was made to produce a hybrid of wheat and rye. French researchers finally produced a viable offspring of the two grains in the late 1930s and named it *triticale* by combining the genus of wheat (*Triticum*) and that of rye

high on rye?

Because of rye's hardiness and ability to grow in damp, cool areas it was a lifesaver to the early colonists in New England. When wheat seeds brought from Europe didn't grow well, Massachusetts settlers soon relied on rye (often mixed with corn) for their bread. About the only problem with the use of rye is that it is occasionally contaminated in the field by a parasitic fungus that produces a complex of poisons called ergot. Because one of these poisons is converted to LSD when baked in bread, it produces hallucinations. It is now thought that the madness of Salem's 1692 witch trials was caused by bread made from ergot-contaminated rye flour. Fortunately, today's rye flour is monitored for the presence of ergot.

(*Secale*). Available as berries, flakes, or flour, triticale is more nutritious than either of its parents.

The "staff of life" for most of Europe and North America, *wheat* is extremely versatile. Because of its high gluten content, it produces superior baked goods and is often added to low-protein grains to improve their baking characteristics. Available as berries, bran, and bulgur as well as a variety of flours, wheat has recently returned to the saucepan in a range of side and main dishes. Semolina, the endosperm of the hardest durum wheat, is ground into a golden flour, which appears in pastas, bread, and couscous.

grain nutrition

The U.S. Department of Health and Human Services, the National Cancer Institute, the American Heart Association, many other health organizations, and the Surgeon General concur that we should reduce the amount of fat and calories and increase the amount of fiber and complex carbohydrates in our diets. Grains are low in fat and calories, high in fiber, and one of the best sources for complex carbohydrates. They're also a cholesterol-free protein source—1 cup generally provides less than 2 grams of fat, no cholesterol, and from 120 to 280 calories. Current guidelines advise that more than 50 percent of the calories you consume should be supplied by grains and other complex carbohydrates.

Grains are rich in both soluble and insoluble fiber—the former is thought to help lower blood cholesterol, the latter to help prevent constipation and protect against some cancers. Whole grains provide signifi-

grain nutrition

General information based on 1/2 cup cooked grain.

	amaranth	barley	buckwheat	corn	millet	oats
Calories	140	87	92	130	119	62
Protein (g.)	5	2	3	4	4	3
Carbohydrate (g.)	25	22	20	29	24	11
Total fat (g.)	2.5	<1	<1	2	1	1
Saturated fat (g.)	<1	-	<1	<1	<1	<1
Cholesterol (mg.)	0	0	0	0	0	0

minerals

	amaranth	barley	buckwheat	corn	millet	oats
Calcium	*	-	-	-	-	-
Copper	✔	✔	-	-	✔	-
Iron	*	**	-	-	*	*
Magnesium	**	*	*	*	*	-
Manganese	✔	-	✔	-	✔	✔
Phosphorus	**	-	-	*	*	*
Potassium	✔	✔	-	-	-	-
Sodium (m.g.)	8	2	4	22	2	1
Zinc	*	*	-	-	-	-

vitamins

	amaranth	barley	buckwheat	corn	millet	oats
B$_6$	*	-	-	-	-	-
C	-	-	-	-	-	-
Folacin	*	-	-	-	*	-
Niacin	-	*	-	-	-	-
Riboflavin	-	*	-	-	-	-
Thiamin	-	**	-	*	-	-
Fiber	NA	H	M	M	M	L

quinoa	rye	triticale	wheat
94	88	110	83
3	4	5	3
17	17	23	19
2	<1	1	<1
<1	<1	<1	<1
0	0	0	0
✔	✔	✔	-
*	**	**	*
*	**	**	-
✔	✔	✔	✔
*	**	**	-
✔	-	✔	-
5	2	2	5
*	**	*	*
*	*	-	-
-	-	-	-
*	*	*	-
*	*	-	-
*	*	-	-
*	*	**	-
N/A	H	H	H

cant sources of vitamins, especially vitamin E and the B vitamins—B_6 (pyroxidine), folacin, niacin, and riboflavin. Grains with the germ removed contain less vitamins and fiber. Only amaranth is a source of vitamin C, providing just 7 percent of the Recommended Dietary Allowance (RDA) in $1/2$ cup of dry seeds. Grains are chock-full of minerals too, providing copper, iron, magnesium, manganese, phosphorus, potassium, and zinc. The kinds and amount of minerals and vitamins vary with the grain (see "Grain Nutrition," pages 6-7). For instance, only amaranth is a good source of calcium.

In general, the protein in true grains is incomplete. Like other plant sources of food, grains are missing or low in the essential amino acid lysine (a part of proteins that the body cannot make). The key to getting the most from grains—as is

the case with beans—is to combine them with other protein sources to make a complete protein, such as White Bean and Barley Soup (page 46), Bulgur Garden Vegetable Salad (page 52), and Red Pepper Polenta with Ratatouille (page 86). By varying your diet to include grains and beans, seeds, nuts, or a small amount of animal food (cheese, eggs, fish, milk, poultry, meat), the available portion is complete, or balanced.

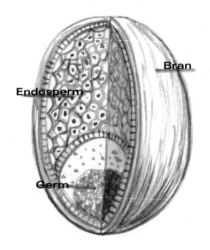

a grain glossary

Amaranth *(Amaranthus):* A tiny mustard-colored seed about the size of a poppy seed with a nutty almost peppery flavor; not a true grain. Available as seeds and flour.

Barley *(Hordeum vulgare):* A true grain with a nutty flavor and a slightly chewy texture. Available hulled, or as flakes, grits, pearl, or pot (Scotch).

popped amaranth

When toasted, amaranth seeds puff like popcorn. Toast a small amount in an ungreased skillet, tossing and stirring them over high heat for a few seconds: 1 tablespoon will yield about 3 to 4 tablespoons popped amaranth. Use immediately. Adds flavor and texture to salads, pancakes, soups, and stews.

Bran: The outer protective layer of the endosperm and germ, which contains most of the B vitamins, minerals, and fiber; it is removed during milling.

Buckwheat *(Fagopyrum esculentum):* These triangular granules have a strong, distinctive nutlike flavor that marries well with earthy and hearty foods. Not a true grain but the fruit of a leafy plant belonging to the same family as sorrel and rhubarb. Available as grits and whole, roasted, and cracked groats (kasha).

Bulgur *(Triticum):* A processed form of cracked wheat produced by steaming and drying whole wheat kernels. This tender grain has a chewy texture and is available in coarse, medium, and fine granulation.

Corn *(Zea mays):* A yellow, white, red, or blue kernel that can be eaten fresh, dried, and cracked, or ground into a variety of products. Originally the generic term for any grain in Europe.

Cornmeal: Ground dried corn; available in coarse, medium, or fine granulation; available yellow, white, red, or blue.

Couscous: Precooked and dried semolina, physically resembling grits; a kind of instant pasta.

Cracked wheat: Whole wheat berries that have been cracked or broken into medium or coarse granules for faster cooking. You can make this at home in a heavy-duty blender by processing 2 cups of wheat berries on high speed for a few minutes.

Endosperm: The largest and starchiest part of the kernel, often more than 80 percent of its weight; it is used to make flour.

Flakes (or rolled): Flattened grains, sometimes steamed before rolled; often used as a cereal.

Flour: Finely ground and sifted meal from the endosperm of grains.

Germ: The smallest part of the grain, which would sprout to form a new plant if planted; the only fat-containing part of the kernel. It is rich in vitamin E and the B vitamins. *Also known as* embryo.

Grain: An edible seed. True grains are members of the grass family.

Gramineae: The family to which the grains belong. Amaranth, quinoa, and buckwheat belong to different botanical families and are botanically not true grains.

Grits: Grains cracked into very small pieces; sometimes toasted.

Groats: Whole crushed kernels that have not been milled, polished, or heat treated. Available as coarse, medium, and fine granulations. Wheat groats are referred to as berries.

Hominy: Corn soaked in a lime water solution to remove the hull, then dried. This process brings out the niacin (a B vitamin), making the corn more nutritious. Available dried or canned.

Hull: A fibrous shell covering the endosperm of the grain.

Husk: The outer inedible fibrous covering of the grain.

Insoluble fiber: The kind of fiber (parts of plants that can't be digested by enzymes in the human digestive tract) that helps prevent constipation and protect against some cancers.

Instant oatmeal: Thinly rolled oats that have been precooked and dehydrated. To prepare, rehydrate with boiling water.

Kamut (Triticum polonicum): Ancient variety of wheat. Available as berries, flour, grits, and pasta.

Kasha: Buckwheat groats, roasted to produce a more intensive nutlike flavor. Available whole and in coarse, medium, and fine granulations.

Kernel: A seed.

Lysine: An essential amino acid that is missing in true grains, thus the protein of grains is incomplete. Combine grains with other protein foods that contain lysine, such as cheese, meat, seeds, beans, nuts, and nongrains (quinoa, amaranth, and buckwheat) to complement the grain's protein.

Maize: The Aztec name for corn.

Malt: The result of a process of soaking a whole grain (usually barley), then sprouting and drying it, during which the bran is converted to enzymes that change the starches to sugar. Traditionally used for alcoholic beverages. Also ground into a fine powder and used for malted vinegar and malted milk.

Masa harina: Flour made from dried hominy.

Maslin: Flour and bread produced in Europe from cultivated wheat and wild rye that grow together in the same field.

Millet: Tiny, round, shiny, pale yellow to reddish orange seeds that have a bland flavor. Because it's gluten-free, it is usually tolerated by those allergic to wheat. Available whole, cracked, and ground.

Oats (Avena sativa): Available as bran, groats, rolled, or steel-cut (Irish).

Pearl barley: Milled six times to remove the outer husk and bran layer before being steamed and polished; chewy texture with nutlike taste.

Polenta: Porridge made from cornmeal.

Posole: Hispanic name for whole hominy. Available dried, canned, and in some ethnic supermarkets, ready to eat.

Quinoa (Chenopodium quinoa): Pronounced *keen'-wah;* tiny ivory-colored, flat-oval seeds about the size of millet that have a delicate flavor. Not a true grain but the fruit of a leafy plant belonging to the same family as spinach and Swiss chard—the leaves of the plant can be cooked and eaten like spinach. Available as whole seed, flour, and pasta.

Rice (Oryza sativa): Cultivars include long, medium, and short grain and regular and aromatic. Available as whole, instant, grits, meal, and flour. (For more information see *Rice* (Harmony, 1991).

Rolled oats: Oats that have been steamed and pressed flat between steel rollers, which shortens cooking time. *Also known as* oat flakes.

Rye (Secale cereale): Similar in appearance and closely related to wheat but contains less gluten. Available cracked and as flakes, grits, berries, and flour.

Saponin: Naturally occurring bitter-tasting glucoside coating on each grain of quinoa; to remove, rinse grain in water.

Scotch barley: Milled three times; some of the bran layer remains intact. It is coarsely ground whole hulled barley.

Semolina: Coarse granular meal made from the endosperm of durum wheat (an amber-colored, hard, high-gluten wheat); generally used to make pasta.

Soluble fiber: The kind of fiber (parts of plants that can't be digested by enzymes in the human digestive tract) that lowers blood cholesterol.

Spelt (Triticum spelta): Ancient variety of wheat. Available as berries, flour, and pasta.

Steel-cut (or Irish) oats: The least processed form of oats;

groats that have been thinly sliced lengthwise. They have a chewy, hearty consistency.

Super grains: Amaranth and quinoa, which are considered "super" special because they're rich in proteins—including lysine, the amino acid missing in true grains.

Teff: The tiniest grain. It has a light, nutty flavor.

Triticale: Pronounced *tri-ti-kay'-lee*. The nutritious hybrid of wheat (*Triticum*) and rye (*Secale*), it contains more protein than wheat, although less converts to gluten when it's used in baking. Available as berries, flakes, and flour.

Wheat (Triticum aestivum, T. durum, and the more primitive T. monococcum and T. dicoccum): One of the oldest cultivated grains. Available as berries, bulgur, cracked, grits, flakes, and flour.

Wheat berries: Whole kernels with hulls removed that have not been milled, polished, or heat treated. They have a nutlike robust flavor with a very chewy texture; soaking or toasting reduces cooking time. *Also known as* groats.

Whole grain: A grain that retains its edible outside layers; one that has not been refined.

Wild rice (Zizania aquatica): The grain of an aquatic grass native to North America; not a true rice.

basic grain recipes

Oats—A grain which in England is generally given to horses, but in Scotland supports the people.

—SAMUEL JOHNSON, A Dictionary of English Language, 1755

basic grain cooking

For the ratio of grain to liquid and the cooking time, refer to "General Grain Cooking" (page 18).

basic simmering: *To cook* amaranth, barley, buckwheat, hominy, millet, posole, quinoa, rye berries, *or* wheat berries, *bring the liquid to a boil over high heat, add the appropriate amount of the grain, stir, cover, reduce the heat to low, and cook until the liquid is absorbed and the grain is tender. Fluff with a fork and serve.*

basic steeping: *To steep* bulgur, couscous, *or* instant oatmeal, *pour boiling liquid over the grain, cover, and let stand until the grain is tender.*

basic pilaf: *To make a pilaf, cook a chopped onion, garlic, and/or other vegetable in oil until softened; add* amaranth, barley, buckwheat, millet, quinoa, rye berries, steel-cut oats, *or* wheat berries, *stir gently until the grain is toasted and coated with the oil, add liquid, bring to a boil over high heat, stir, cover, reduce the heat to low, and cook until the liquid is absorbed and the grain is tender. Fluff with a fork and serve.*

basic cereal: *To make cereal, combine cold water and* meal, grits, *or* flakes *of all grains. Bring to a boil over high heat, stirring constantly; cover; reduce heat to low; and cook to desired consistency.*

note: *When simmered, amaranth becomes a gelatinous textured mush. To make it more appealing, we suggest cooking a small portion of it with another grain; follow the cooking time for that grain.*

The classic method of preventing buckwheat from turning to mush is to coat it with egg, then sauté it until the egg dries. This gives the kernels a protective seal that will keep them separate during cooking.

Unlike other grains, when quinoa is done, the grains appear translucent and a white crescent-shaped germ is visible.

tips for perfect grains

- *Always measure the grain and liquid accurately and cook it following recipe instructions.*
- *Cooking times for the same kind of grain will vary, depending on the length of time since harvest and the storage conditions. Check for tenderness about 5 minutes before the shortest cooking time and don't be surprised if a grain sometimes takes a bit longer than the longest listed cooking time.*
- *Whole grains should be sorted and rinsed before using. Processed grains can be used directly from the package.*
- *Toasting before cooking can reduce the cooking time of hulled barley, wheat berries, and steel-cut (Irish) oats by as much as 20 minutes. Either spread the grain out on a rimmed baking sheet and toast 5 minutes in a 400° F. oven, or place in a large, ungreased skillet and stir over medium heat 5 minutes or until the grain starts to look toasted.*
- *Soaking whole grains for several hours or overnight in the refrigerator will reduce cooking time.*
- *Flour can be made from any rolled grain in a food processor fitted with a steel blade. Use 1 cup of the rolled grain at a time and process until the desired fineness has been reached. We don't suggest making flour from whole grains in the food processor, but there are several good grain mills on the market that are made for that purpose.*

general grain cooking

GRAIN	Grain Product	Liquid/ 1 cup	Approximate Yield in Cups	Preparation Time
Amaranth		3	$2^2/_3$	15 min.
Barley				
	Hulled	4	$2^1/_2$	1 hour 45 min.
	Pearl	3	4	45 to 55 min.
	Quick	2	3	10 to 12 min.
	Grits	$^2/_3$	$^2/_3$	2 to 3 min. (plus 5 min. standing)
	Flakes	3	2	30 min.
Buckwheat				
	Groats			
	unroasted	2	$3^1/_2$	15 min.
	roasted	2	3	15 min.
	Kasha			
	fine	2	4	8 to 10 min.
	medium	2	4	8 to 10 min.
	coarse	2	4	10 to 12 min. (plus 5 min. standing)
Corn				
	Cornmeal	4	$3^1/_2$	10 min.
	Hominy	5	3	5 to 6 hours

GRAIN	Grain Product	Liquid/ 1 cup	Approximate Yield in Cups	Preparation Time
Millet		3	5	25 to 30 min.
Oats				
	Steel-cut (Irish)	4	2	20 min.
	Old-fashioned rolled	2	2	5 min.
	Instant	2	2	1 min. (plus 3 to 5 min. standing)
Quinoa		2	4	15 min.
Rye				
	Berries	3	3	1 hour 55 min.
	Flakes	3	2^1/$_2$	1 hour 5 min.
Triticale				
	Berries	3	2^1/$_2$	1 hour 45 min.
Wheat				
	Berries	3	2^1/$_2$	1 hour 10 min.
	Couscous, quick cooking	1^1/$_2$	3	5 to 10 min.
	Cracked	2	2	15 min.
	Flakes	4	2	50 to 55 min.
	Bulgur			
	fine	2	3	15 min.
	medium	2	3	15 min.

using leftover grains

- *Leftover grains keep well in the refrigerator for 1 to 2 days or in the freezer (except hominy, which gets mushy with freezing) for up to 6 months.*
- *Package leftover grains for freezing in 1- or 2-cup portions.*
- *To reheat frozen cooked grains, add 2 tablespoons water and cook over low heat until heated through, or microwave at full power (100 percent) for 2 minutes. Or put them directly into boiling broth or soup and heat the soup until it returns to boiling.*
- *Freeze cooked breakfast cereals in $^1/_2$- to $^3/_4$-cup portions. The night before you want to use them, unwrap the frozen cereal and place in a bowl. Cover and let it thaw overnight in the refrigerator; then microwave it at full power (100 percent) for $1^1/_2$ minutes.*

storing grains

It is best to store all grain products in tightly sealed containers and in a cool, dry location. Processed products keep better than natural, whole-grain products. Because of their natural oil content, whole grains tend to become rancid and attract insects, particularly in warm climates. We recommend storing unprocessed grains in the refrigerator for 3 to 4 weeks or in the freezer for up to 6 months.

breakfast

On either side of the

river lie

Long fields of barley

and of rye,

That clothe the wold

and meet the sky;

And through the field

the road runs by

To many-towered

Camelot.

—ALFRED,

LORD TENNYSON, *The*

Lady of Shalott, 1832

buckwheat waffles

6 servings

1. Combine the flours, sugar, baking powder, and salt in a medium bowl.
2. Separate the eggs; place the whites in a small, deep bowl and the yolks in a custard cup.
3. Beat the egg whites until they are stiff but not dry.
4. Combine the milk, oil, and egg yolks in a 2-cup glass measuring cup or another small bowl.
5. Preheat the waffle iron. Add the milk mixture to the buckwheat mixture and stir just until the dry ingredients are moistened. Carefully fold in the beaten whites.
6. Bake waffles according to waffle iron manufacturer's directions. Serve immediately.

¾ cup buckwheat flour

¾ cup all-purpose flour

1 tablespoon sugar

1 tablespoon baking powder

¼ teaspoon salt

2 eggs

1½ cups milk

2 tablespoons vegetable oil

buttermilk oatmeal pancakes

An unusual variation on traditional pancakes. Try them with warm maple syrup.

about 1 dozen 3-inch pancakes (4 servings)

1. Combine the oats and buttermilk in a medium bowl; let stand 15 minutes.

2. Add the flour, sugar, baking powder, baking soda, and salt. Mix together the water, egg, maple syrup, and butter; add to flour mixture. Stir until blended.

3. Heat 1 teaspoon oil in large skillet or griddle. Pour $1/4$ cup batter onto hot surface for each pancake. Fry until golden on one side. Turn and fry on other side until golden and cooked through. Serve immediately.

1 cup old-fashioned rolled oats

1 cup buttermilk

1 cup all-purpose flour

2 tablespoons sugar

1 teaspoon baking powder

1 teaspoon baking soda

½ teaspoon salt

¼ cup water

1 egg, lightly beaten

2 tablespoons maple syrup

2 tablespoons melted butter or vegetable oil

Vegetable oil for frying

corn dodgers

These are like an American version of the English muffin or crumpet.

12 servings

1. Combine the cornmeal, flours, baking powder, and salt in a medium bowl. Stir in the milk, egg, butter, and maple syrup until a soft dough forms.

2. Heat 1 teaspoon oil in a large skillet, griddle, or electric frying pan over low heat. Spoon the dough into the skillet a scant $1/4$ cup at a time. Spread to make a 3-inch round.

3. Fry until golden on one side. Turn and fry on other side until golden and cooked through. Split crosswise and serve immediately with butter and jam.

$1^{2}/_{3}$ cups yellow cornmeal

$1/_{3}$ cup rye flour

$1/_{3}$ cup whole wheat flour

1 tablespoon baking powder

$1/_{2}$ teaspoon salt

1 cup milk

1 egg

3 tablespoons melted butter

2 tablespoons maple syrup

Vegetable oil for frying

Butter and jam

granola

Serve as a snack or cereal. Or use as a topping on yogurt, ice cream, or fruit.

about 6 cups

1. Preheat oven to 300° F. Combine the oats (or other flakes), coconut, almonds and sesame seeds if desired, wheat germ, and salt in a large bowl. Combine oil, honey, cinnamon, zest, and vanilla in a small bowl; pour over oat mixture and toss until evenly moistened.

2. Spread the mixture evenly on two rimmed baking sheets and bake, stirring frequently, until the oats are golden brown, 15 to 20 minutes. Remove from the oven, stir in the currants, and cool to room temperature. Store in an airtight container.

4 cups old-fashioned rolled oats, rye flakes, or barley flakes

½ cup unsweetened dried coconut

½ cup sliced almonds (optional)

¼ cup sesame seeds (optional)

¼ cup wheat germ or wheat bran

½ teaspoon salt

¼ cup vegetable oil

¼ cup honey or maple syrup

1 tablespoon ground cinnamon

2 teaspoons grated orange zest

1½ teaspoons vanilla extract

1 cup dried currants

lacy cornmeal pancakes

To get the traditional whispy edges and lacy surface, the pan must be hot and well greased.

about 8 servings (24 pancakes)

1. Combine the cornmeal, flour, sugar, baking soda, and salt in a medium bowl. Stir in buttermilk, egg, and butter just until dry ingredients are moistened.

2. Heat 1 teaspoon oil in large skillet, griddle, or electric frying pan. Spoon batter into skillet to make a 3-inch round. Fry until golden on one side. Turn and fry on other side until golden and cooked through. Serve immediately.

1½ cups cornmeal

1 tablespoon all-purpose flour

2 tablespoons sugar

1 teaspoon baking soda

¼ teaspoon salt

2 cups buttermilk

1 egg

2 tablespoons melted butter

Vegetable oil for frying

muesli

Muesli means mush in Swiss. This breakfast cereal was developed near the end of the nineteenth century by Swiss physician Dr. Bircher-Benner.

4 servings

1. The day before serving, combine the oats, rye, 1¼ cup yogurt, milk, dried fruit, zest, and cinnamon in a large bowl. Stir well, cover, and refrigerate overnight.

2. Just before serving, stir in the apple and almonds. If necessary, add more yogurt to thin.

1 cup rolled oats

½ cup rye flakes

1¼ to 1½ cups nonfat plain yogurt

¾ cup skim milk

½ cup chopped prunes, dried currants, dates, or apricots

1 teaspoon grated lemon zest

½ teaspoon ground cinnamon

1 small apple, peeled and grated

½ cup chopped almonds

blueberry corn muffins

12 muffins

1. Preheat the oven to 375° F. Line a 12-cup muffin tin with paper baking cups.
2. Combine cornmeal, flour, 1/3 cup sugar, baking powder, baking soda, zest, and salt in a medium bowl. Combine eggs, yogurt, vanilla, and butter in a small bowl.
3. Add the yogurt mixture to the cornmeal mixture and stir just until combined. Do not overmix; the batter should be lumpy. Stir in the blueberries.
4. Divide batter evenly among the muffin cups. Sprinkle with the cinnamon sugar. Fill any empty cups with water. Bake for 20 to 25 minutes or until the centers spring back when gently pressed. Remove the muffins from the pan to a wire rack and let cool.

1 1/4 cups yellow cornmeal

3/4 cup all-purpose flour

1/3 cup sugar

1 tablespoon baking powder

1/2 teaspoon baking soda

2 teaspoons grated lemon zest

1/2 teaspoon salt

2 eggs

1 cup plain yogurt or sour cream

1 teaspoon vanilla extract

4 tablespoons butter, melted

1 cup blueberries

1 tablespoon sugar mixed with 1/4 teaspoon ground cinnamon

soups and starters

The corn is as high as an elephant's eye, An' it looks like it's climbin' clear up to the sky.

—Oscar Hammerstein II, *Oh, What a Beautiful Mornin'*, 1943

barley and bell pepper cakes

An unusual side dish, this also works well as an appetizer. Serve as an accompaniment to roast Cornish hen, chicken, or turkey.

about 9 servings (18 cakes)

1. Combine the barley, bell peppers, prosciutto, scallions, eggs, and garlic in a medium bowl. Then stir in the flour, salt, and black pepper. Shape 1/4 cup barley mixture into a 2-inch flattened cake. Repeat to make 18 cakes. Refrigerate for 1 hour or until firm.

2. Heat 1 teaspoon oil in a large skillet over medium-high heat. Fry cakes on both sides until crisp and lightly browned, adding more oil as needed, 5 to 7 minutes.

3 cups hot cooked barley (see pages 16, 18)

2 medium red bell peppers, cored, seeded, and diced (about 1 1/2 cups)

1/4 cup finely minced prosciutto or other smoked ham

3 scallions, minced (about 1/2 cup)

2 eggs

1 garlic clove, minced

1/2 cup all-purpose flour

Salt and freshly ground black pepper to taste

Vegetable oil for frying

buckwheat blini

7 to 8 servings

1. Preheat the oven to 250° F. Combine the flours, sugar, yeast, and salt in a medium bowl.

2. Stir the water and 1 tablespoon oil into the flour mixture until a smooth, thick batter forms. Set the batter aside until it bubbles, 15 to 20 minutes.

3. Heat 1 teaspoon oil over medium heat in a large skillet. Drop the batter into the skillet, 1 level measuring tablespoon at a time, to make blini about 2 inches in diameter. Fry until lightly browned on one side; turn and fry until browned on the other, 3 to 5 minutes in all. Repeat to make 20 to 24 blini.

4. Keep blini warm in the oven while frying the others.

5. To serve, divide the blini onto individual serving plates. Top each blini with a dollop of sour cream, a few strips of smoked salmon, and, if desired, a small sprig of dill.

1 cup buckwheat flour

1 cup all-purpose flour

2 teaspoons sugar

1 teaspoon (about ½ package) fast-rising yeast

½ teaspoon salt

1 cup hot water (120° to 125° F.)

Vegetable oil

1 cup sour cream

4 ounces thinly sliced, cold, smoked salmon, cut into julienne strips

Fresh dill (optional)

beef and barley soup

Bonnie's sister Meredith Spangenberg inspired this hearty soup, a favorite at her annual winter soup party.

8 to 10 servings

1. Heat the oil in a 4-quart stock pot or saucepan over medium-high heat. Add the onions and garlic and cook over medium heat until the onions begin to caramelize, about 10 minutes. Add the celery and carrots; cook until vegetables soften, about 5 minutes. Push the vegetables to one side of the pot, add the beef, and brown both sides 3 to 5 minutes.

2. Add the water, $1/2$ cup parsley, salt, and pepper. Bring to a boil over high heat, reduce to low, cover, and simmer until the beef is tender, at least 1 hour. Remove the beef and let cool. Remove the meat from the bone and the marrow from inside the bone, chop into bite-size pieces, and return to the pot along with the barley.

3. Cook, covered, for 45 minutes or until barley is tender. Taste and adjust seasoning. Serve sprinkled with the remaining parsley.

1 tablespoon oil

1½ large onions, chopped (about 3 cups)

4 cloves minced garlic

4 ribs celery, chopped (about 3 cups)

1 pound carrots, peeled and chunked

2½ pounds bone-in beef shank

8 cups water

¾ cup chopped fresh parsley

Salt and freshly ground black pepper, to taste

1 cup pearl barley, rinsed and drained

corn, millet, and crab chowder

4 servings

1. Sauté the bacon over medium heat in a heavy, 5-quart saucepan or Dutch oven until fat begins to collect in the pan, about 3 minutes. Add the onion and bell pepper and sauté until the vegetables are lightly browned.

2. Add the water and corn kernels to the onion and bell pepper in the saucepan. Bring the mixture to a boil over medium heat, stirring occasionally.

3. Combine the half-and-half and the cornstarch in a small bowl. Whisk the mixture into the simmering soup along with the salt. Cook, stirring constantly, until the mixture boils and thickens slightly; cook 2 minutes longer.

4. Stir the millet, crabmeat, parsley, and salt into the simmering mixture; return just to boiling and serve.

1 slice bacon cut into ¼-inch crosswise strips

1 small onion, finely chopped (about ½ cup)

2 tablespoons finely chopped red bell pepper

3 cups water

1½ cups fresh or frozen corn kernels

1 cup half-and-half or milk

2 teaspoons cornstarch

½ cup cooked millet (see pages 16, 18)

½ cup fresh crabmeat

2 tablespoons chopped fresh parsley

¼ teaspoon salt

hummus with quinoa

Adding quinoa to this traditional Mideastern dish makes the hummus lighter. Serve as a sandwich spread or a dip for fresh vegetables, crackers, or pita bread.

4 to 6 servings

1. In a food processor or blender, puree the garbanzos, quinoa, lemon juice, tahini, garlic, cumin seeds, and salt. Refrigerate for at least 2 hours to allow the flavors to blend.
2. Stir in the parsley. Taste and adjust seasoning.

variation: *Thin with additional lemon juice and use as a sauce on falafel or steamed vegetables.*

1 cup cooked or canned garbanzos (chick-peas), drained

1 cup cooked quinoa (see pages 16, 18)

1/4 cup freshly squeezed lemon juice

2 tablespoons tahini (sesame seed paste) or good-quality olive oil

2 garlic cloves

1/4 teaspoon cumin seeds

1/4 teaspoon salt

1/3 cup chopped fresh parsley

corn, quinoa, and red pepper chili chowder

Serve with crusty bread or tortilla chips. If you prefer a milder chowder, just omit the jalapeños.

4 to 6 servings

1. Heat the butter in a 4-quart saucepan over medium heat. Add the onion, bell pepper, jalapeño peppers, and quinoa; sauté until onion is softened and quinoa is toasted and aromatic, about 5 minutes.

2. Add the broth, potatoes, thyme, salt, cayenne (if desired), and bay leaf. Bring to a boil, reduce heat, cover, and simmer over low heat for 20 minutes.

3. Add the milk, corn, and black pepper. Simmer an additional 5 to 7 minutes, until warmed through. Remove and discard the bay leaf. Taste and adjust seasoning; stir in parsley and serve.

1 tablespoon butter

1 large red onion, chopped (about 1½ cups)

2 medium red bell peppers, cored, seeded, and chopped (about 1½ cups)

2 jalapeño peppers, seeded and minced

½ cup quinoa, rinsed in cold water and drained

4 cups chicken or vegetable broth

3 potatoes (about 1 pound), peeled and cut into ½-inch cubes

1 teaspoon dried thyme

1 teaspoon salt

¼ teaspoon cayenne (ground red) pepper (optional)

1 bay leaf

2 cups milk or half-and-half

2 cups fresh or frozen corn kernels

Freshly ground black pepper to taste

½ cup chopped fresh parsley

variations: *Substitute corn on the cob, when available. Slice 2 ears fresh shucked corn, crosswise, into 3/4-inch rounds. To make into a hearty entree, add 1 pound of sliced chorizo or other garlic sausage along with the onion.*

note: *Be sure to wash your hands after handling jalapeño peppers. The pepper's volatile oils could burn your skin or eyes.*

fresh corn and amaranth tamales

4 servings

1. Combine the *masa harina*, water, and butter in a large bowl, stirring vigorously. Fold in the corn, amaranth, cheese, bell peppers, salt, and Tabasco sauce until well blended.

2. Bring several inches of water to a boil in the bottom of a steamer or in a large kettle.

3. To make tamales, grease corn husks or aluminum foil rectangles. Divide the *masa harina* mixture onto husks or foil, forming each into a 4 × $^3/_4$-inch log. If using husks, wrap husks loosely over *masa harina* mixture and tie ends securely with string. If using foil, roll mixture in foil and twist ends.

4. Stack tamales in basket of steamer or on rack over water in large kettle. Steam tamales 1 hour, or until *masa harina* has cooked through.

5. Remove to serving platter and serve with salsa for dipping, or unwrap each tamale and place on a serving plate with a bowl of salsa.

1 cup *masa harina* or corn flour

1 cup boiling water

1 tablespoon butter

1 cup fresh corn kernels

2 tablespoons cooked amaranth (see pages 16, 18)

½ cup shredded Monterey Jack cheese

¼ cup finely chopped red bell pepper

¼ cup finely chopped green bell pepper

¼ teaspoon salt

2 to 4 drops Tabasco sauce or other hot red-pepper sauce

Fresh inner corn husks, blanched, or aluminum foi

tabbouleh

Although this is traditionally made with mint, we chose to use basil—a more popular member of the mint family—for flavoring. Serve as a side dish, over a salad, or in pita bread.

10 to 12 servings

3½ cups boiling water

2 cups bulgur, medium granulation

1½ cups chopped fresh parsley

⅓ cup olive oil

⅓ cup freshly squeezed lemon juice

¼ cup chopped fresh basil

2 garlic cloves, minced

1 teaspoon or more salt

¼ teaspoon cayenne (ground red) pepper

¼ teaspoon ground cumin

Freshly ground black pepper

2 cucumbers peeled, seeded, and diced (about 2 cups)

4 scallions, including green tops, minced (about ⅔ cup)

3 large ripe tomatoes, peeled, seeded, and diced (about 3 cups)

1. Pour the water over the bulgur in a large bowl. Cover and let stand until bulgur has doubled in bulk and most of the liquid has been absorbed, about 30 minutes. Drain in colander to remove excess water.

2. Whisk together parsley, oil, lemon juice, basil, garlic, salt, cayenne, cumin, and pepper in a medium bowl. Mix in cucumbers and scallions. Pour over cooled bulgur, toss well, and let stand at room temperature for at least 1 hour to allow flavors to blend.

3. Taste and adjust seasoning. Mix in tomatoes just before serving.

hungarian goulash soup with barley and rice

Although noodles are classic with goulash, we took the liberty of including a few grains in this hearty soup.

8 to 10 servings

1. Fry the bacon in a 6-quart saucepan over medium heat until crisp. Remove to paper toweling to drain. Add onions and cook until golden, about 10 minutes.
2. Add the beef, stirring frequently, until browned, about 10 minutes. Add the tomato paste, garlic, paprika, caraway seeds, salt, and pepper; cook, stirring 1 minute. Add the water, bring to a boil, reduce heat, cover, and simmer for 45 minutes.
3. Add the rice and barley; stir, cover, and simmer an additional 45 minutes. Taste and adjust seasoning. Garnish with the crumbled bacon.

vegetarian variation:
Omit the bacon and beef. Cook vegetables in 1 tablespoon oil.

3 slices bacon
2 large onions, minced (about 3 cups)
1½ pounds beef chuck or round, cut into ¾-inch cubes
¼ cup tomato paste
2 garlic cloves, minced
2 tablespoons Hungarian paprika
1 tablespoon caraway seeds, crushed
2 teaspoons salt or to taste
Freshly ground black pepper to taste
2 quarts boiling water
½ cup brown rice
½ cup pearl barley, rinsed and drained

scrapple

This Pennsylvania Dutch breakfast specialty used to be a tasty way of using some of the scraps when butchering pork. We like it anytime.

4 servings (12 slices)

4 ounces ground pork or mild pork sausage (see note)

½ teaspoon salt

½ teaspoon cracked black pepper

¼ teaspoon rubbed sage (optional)

3 cups water

1 cup white cornmeal

Vegetable oil

Honey or golden syrup

1. Sauté the pork, salt, pepper, and sage, if desired, in a heavy 3-quart saucepan over medium heat, breaking up the pork into small pieces—about 8 minutes. When the pork is cooked through, add the water and cornmeal, stirring to loosen browned-on bits. Bring to a boil, stirring constantly. Reduce the heat to low, cover, and cook, stirring occasionally, for 10 minutes.

2. Meanwhile, generously coat the inside of an 8 × 4½-inch loaf pan with oil. Turn the cooked cornmeal mixture into the oiled pan and set aside to cool to room temperature. Once cool, cover tightly and refrigerate until firm, 4 to 6 hours.

3. To serve, unmold the scrapple onto a cutting board and cut crosswise into 12 slices. Heat 1 tablespoon vegetable oil over medium heat in a large skillet and fry the scrapple until golden brown on both sides and

heated through. Serve with honey or
golden syrup.

note: *If using sausage, reduce the
salt and pepper when cooking by
half. Once the scrapple is fully
cooked, add salt and pepper to taste.*

sweet-and-sour rye meatballs

4 servings

1. Combine the beef, pork, rye flakes, $1/3$ cup apple juice, the egg, 2 tablespoons onion, 2 tablespoons parsley, $1/2$ teaspoon salt, the black pepper, and allspice in a large bowl.

2. Place the flour in a small bowl. Drop the meat mixture by level measuring tablespoonsful into the flour and roll until the meatballs are lightly coated. Reserve any remaining flour.

3. Heat half of the butter in a large skillet over medium heat. Add half the meatballs and sauté until they are browned on all sides, about 5 minutes. Remove to a bowl and repeat with the remaining meatballs and butter.

4. Add the remaining onion to the drippings in the skillet and sauté until browned, 1 to 2 minutes. Add $1^{1}/3$ cups more of the remaining apple juice and heat the mixture to a boil, stirring to remove any browned-on bits.

½ pound lean ground beef

½ pound lean ground pork

½ cup rye flakes

2 cups apple juice or cider

1 egg

1 small onion, finely chopped (about ½ cup)

¼ cup chopped fresh parsley

¾ teaspoon salt

¼ teaspoon ground black pepper

¼ teaspoon ground allspice

½ cup rye flour

1 tablespoon butter

1 teaspoon cider vinegar

1 large Golden Delicious or Fiji apple, sliced ¼-inch thick

5. Stir 2 tablespoons of the flour remaining in the bowl (or use additional flour, if necessary) into the remaining $1/3$ cup juice. Add to the mixture in the skillet and cook, stirring constantly, until a thickened sauce forms, about 3 minutes.

6. Stir the remaining $1/4$-teaspoon salt, the vinegar, meatballs, and apples into the sauce in the skillet. Cover; reduce heat to low and cook gently until meatballs are cooked through and apple is just tender, about 10 minutes. Turn meatballs, sauce, and apples into a serving bowl or chafing dish. Top with remaining parsley; provide toothpicks for easy eating.

wild mushroom and barley soup

This delicate, elegant soup is an example of the perfect marriage between grains and mushrooms. If fresh mushrooms are not available, you can use dried mushrooms as a substitute—see variation below.

6 servings

1. Rinse the mushrooms and dry with paper towels. Thinly slice them.
2. In a 6-quart saucepan, heat olive oil over medium-high heat. Add mushrooms and stir until they are soft and have released their moisture, about 5 minutes. Add the shallots, garlic, parsley, thyme, salt, and pepper.
3. Add the broth, bring to a boil over high heat. Stir in the barley; reduce heat to low, cover, and simmer until the barley is tender, about 45 minutes. Taste and adjust seasoning. Add sherry and serve.

1 pound fresh cultivated exotic (wild) mushrooms, such as porcini, portobellos, and/or shiitakes

2 tablespoons olive oil

2 shallots, finely chopped

2 garlic cloves, finely chopped

2 tablespoons chopped fresh parsley

1 tablespoon fresh or 1 teaspoon dried thyme

Salt and freshly ground black pepper to taste

6 cups chicken or vegetable broth

½ cup pearl barley, rinsed and drained

2 tablespoons dry sherry

variation: *Substitute $1/4$ pound good-quality dried porcini or shiitake mushrooms. Soak in warm water to cover until soft, 30 minutes to 2 hours, depending on the mushroom. Drain and slice mushrooms. Strain and save the soaking liquid; add as part of the broth. For a quick soup, add 2 cups cooked barley.*

white bean and barley soup

This hearty soup seems to stick to your ribs. Perfect for a chilly fall or winter day.

6 to 8 servings

1. Pierce the sausage with a fork. Cook in boiling water 5 minutes, drain, and slice into 1/4-inch rounds.
2. Cook the sausages in a 4-quart saucepan over medium-high heat until lightly browned, about 5 minutes. Add the onion and cook until golden, about 5 minutes. Add the celery, carrots, bell pepper, and garlic. Cook 5 minutes, stirring constantly. Add the water, tomatoes, savory, celery seed, salt, and black pepper.
3. Bring to a boil, stir in the barley, reduce the heat to low, and cook until the barley is tender, about 45 minutes. Stir in the beans and cook an additional 5 minutes until warmed through. Taste and adjust seasoning.

vegetarian variation: *Omit the sausage. Cook the vegetables in 1 tablespoon oil until onion is golden. Continue with recipe.*

1/2 pound sweet sausage

1/2 large onion, minced (about 3/4 cup)

2 ribs celery, minced (about 1 1/3 cups)

2 carrots, diced (about 1 cup)

1 medium green bell pepper, cored, seeded, and diced (about 3/4 cup)

2 garlic cloves, minced

4 cups water

1 cup canned crushed tomatoes

2 teaspoons dry savory

1 teaspoon celery seed

Salt and freshly ground black pepper to taste

1/2 cup pearl barley, rinsed and drained

2 cups cooked or canned white beans, drained

salads

Only reapers, reap-

ing early

In among the bearded

barley,

Hear a song that

echoes cheerly

From the river wind-

ing clearly,

Down to towered

Camelot.

—ALFRED,
LORD TENNYSON, *The
Lady of Shalott,* 1832

athenian couscous salad

A light refreshing salad.

4 to 6 servings

1. Pour the broth over the couscous in a medium bowl; stir and cover. Set aside, stirring occasionally, until all the broth has been absorbed, 5 to 10 minutes. Fluff with a fork.

2. Meanwhile, whisk together the parsley, oil, lemon juice, oregano, salt, and pepper in a medium bowl. Add the tomatoes and feta cheese, stir, and let stand until couscous is ready. Add to couscous, mixing well. Taste and adjust seasoning.

3. Serve or chill until 30 minutes before serving. Garnish with sliced cucumbers.

1½ cups boiling chicken or vegetable broth

1 cup couscous

½ cup chopped fresh parsley

6 tablespoons olive oil

¼ cup freshly squeezed lemon juice

2 teaspoons dried or 2 tablespoons fresh oregano

Salt and freshly ground pepper to taste

2 large tomatoes, peeled, seeded, and chopped (about 2 cups)

½ pound feta cheese, crumbled

Sliced cucumbers for garnish

barley and tricolored pepper salad

8 servings

1. Combine the water and barley in a 2-quart saucepan. Bring to a boil over high heat. Reduce heat to low, cover pan, and cook 45 to 60 minutes, or until barley is tender and has absorbed all the liquid.

2. Combine the warm barley and the vinegar, sugar, dill, olive oil, and salt in a large bowl. Cool to room temperature. Stir in bell peppers and scallions. Cover and refrigerate at least 2 hours before serving.

4 cups water

1½ cups pearl barley, rinsed and drained

½ cup cider vinegar

3 tablespoons sugar

3 tablespoons snipped fresh dill

2 tablespoons olive oil

1 teaspoon salt

½ cup *each* chopped green, red, and yellow bell peppers

2 scallions, finely chopped (about ¼ cup)

black bean, red pepper, corn, and quinoa salad

4 servings

1. Heat 1 tablespoon oil in a 1-quart saucepan over medium heat. Add quinoa and stir until toasted and aromatic, about 5 minutes. Stir in broth, cumin, and salt; bring to a boil, reduce heat, cover, and simmer 15 minutes until the liquid is absorbed. Remove from the heat and allow to stand covered for 5 minutes; fluff with a fork.

2. While the quinoa is cooking, whisk together the remaining olive oil, the lime juice, and black pepper in a medium bowl. Stir in the black beans, corn, red pepper, tomato, scallions, cilantro, and parsley. Stir in the quinoa.

3. Taste and adjust seasoning. Serve at room temperature or refrigerate until 30 minutes before serving. Serve on a bed of lettuce.

variation: *For a more pronounced Southwestern flavor, substitute 1 cup of your favorite salsa for the tomato.*

5 tablespoons olive oil

½ cup quinoa, rinsed in cold water and drained

1 cup chicken or vegetable broth

¼ teaspoon ground cumin

¼ teaspoon salt

2 tablespoons freshly squeezed lime juice

Freshly ground black pepper to taste

1 cup cooked or canned black beans, drained

1 cup cooked corn kernels

1 small red bell pepper, cored, seeded, and chopped (about ½ cup)

1 ripe tomato, peeled, seeded, and diced (about 1 cup)

2 scallions, finely chopped (about ¼ cup)

3 tablespoons chopped fresh cilantro

2 tablespoons chopped fresh parsley

Fresh green leaf lettuce or spinach leaves

bulgur and chicken salad

6 servings

1. Peel the cucumber; cut it crosswise into $1/4$-inch thick slices. Cut the slices into quarters. Combine the cucumber slices and the salt in a colander over a small bowl. Set aside 30 minutes.

2. Sauté the chicken in the oil in a large skillet over medium heat until cooked through and lightly browned—about 15 minutes.

3. Remove the skillet from heat. Transfer the chicken to a large bowl.

4. Add the water, 1 tablespoon vinegar, and 1 teaspoon sugar to the skillet; stir to loosen the browned-on bits. Stir in bulgur and set aside 15 minutes.

5. Meanwhile stir the yogurt, remaining 2 tablespoons vinegar and 2 teaspoons sugar, and the dill into the cucumber juice in the small bowl. Cover and refrigerate until served.

6. Add the bulgur mixture, the bell pepper, peas, and cucumbers to the chicken. Cover and refrigerate at least 2 hours before serving.

7. To serve, spoon the bulgur mixture onto a lettuce-lined platter. Top with the yogurt dressing.

1 cucumber

$1/2$ teaspoon salt

2 whole chicken breasts, skinned and boned, and cut into 1-inch pieces (about 1 pound)

1 teaspoon olive oil

1 cup water

3 tablespoons cider vinegar

1 tablespoon sugar

1 cup bulgur, medium granulation

1 (8-ounce) container plain nonfat yogurt

1 tablespoon snipped fresh dill or $1/2$ teaspoon dried dillweed

1 medium red bell pepper, chopped (about $3/4$ cup)

1 cup cooked fresh or frozen green peas

Leaf lettuce to line platter

bulgur garden vegetable salad

An addictive refreshing garden salad.

4 to 6 servings

1. Heat 1 tablespoon oil in a medium saucepan over medium-high heat. Add the bulgur and stir constantly until toasted, about 2 minutes. Add the broth, bring to a boil over high heat, reduce heat to low, cover, and cook for 10 minutes or until liquid is absorbed. Set aside to cool. Fluff with a fork.

2. Whisk together the remaining oil, the vinegar, cilantro, parsley, salt, and black pepper in a large bowl. Add the bell peppers, zucchini, scallions, tomatoes, carrot, and radishes. Toss in the bulgur. Taste and adjust seasoning; serve topped with feta cheese.

¼ cup olive oil

1 cup bulgur, medium granulation

1¾ cups chicken or vegetable broth

¼ cup balsamic vinegar

½ cup finely chopped fresh cilantro

½ cup finely chopped fresh parsley

Salt and freshly ground black pepper to taste

1 medium red bell pepper, cored, seeded, and diced (about ¾ cup)

1 medium yellow bell pepper, cored, seeded, and diced (about ¾ cup)

1 medium zucchini, diced (about 1 cup)

4 scallions, including green tops, minced (about ⅔ cup)

2 large ripe tomatoes, peeled, seeded, and diced (about 2 cups)

1 carrot finely diced (about ½ cup)

3 radishes, finely chopped (about ¼ cup)

¼ pound feta cheese, crumbled

citrus couscous salad

6 to 8 servings

1. In a medium bowl pour broth and orange juice over couscous; stir to combine. Cover and set aside, stirring occasionally, until all the broth has been absorbed, 5 to 10 minutes. Fluff with a fork.

2. Meanwhile whisk together the oil, vinegar, lemon juice, scallions, cilantro, soy sauce, zest, ginger, salt, and pepper in a small bowl. Add to the couscous, mixing well. Taste and adjust seasoning.

3. Serve or chill until 30 minutes before serving. Mix in oranges and pine nuts just before serving.

1 cup chicken or vegetable broth, boiling

½ cup orange juice

1 cup couscous

6 tablespoons olive oil

3 tablespoons white vinegar

3 tablespoons freshly squeezed lemon juice

¼ cup chopped scallions (about 2 scallions)

¼ cup chopped fresh cilantro

2 teaspoons soy sauce

2 teaspoons grated orange zest

2 teaspoons chopped fresh ginger

Salt and freshly ground black pepper to taste

2 oranges, peeled and cut into segments

¼ cup pine nuts, toasted

smoked turkey, herbed barley, and vegetable salad

Our friend Linda Guica, food editor of the Hartford (Connecticut) Courant, *shared this recipe with us. It's great to make ahead to serve as a one-dish luncheon entree along with some hearty whole-grain bread.*

6 to 8 servings

1. Combine the broth, barley, onion, lemon juice, garlic, 1 teaspoon salt, and $1/2$ teaspoon pepper in a 2-quart saucepan over high heat. Bring to a boil, reduce heat to low, cover, and cook 45 to 60 minutes, until barley is tender and the liquid is absorbed. Cool to room temperature in a bowl.

2. Whisk together the olive oil, vinegar, mustard, parsley, dill, chives, salt, and pepper in a small bowl. Pour half over the cooled barley and mix well. Mix in the turkey, carrots, green beans, mushrooms,

3 cups chicken or vegetable broth

1 cup pearl barley, rinsed and drained

$1/2$ small onion, chopped (about $1/4$ cup)

2 tablespoons freshly squeezed lemon juice

1 garlic clove, minced

Salt and freshly ground black pepper

$3/4$ cup olive oil

$1/4$ cup red or white wine vinegar

$1 1/2$ teaspoons Dijon mustard

$1/4$ cup chopped fresh parsley

2 tablespoons snipped fresh dill or 1 teaspoon dried dillweed

1 tablespoon chopped chives

4 ounces thickly sliced smoked turkey, cubed

radishes, and scallions. Add the remaining dressing, toss again, and taste and adjust seasoning. Refrigerate until 30 minutes before serving.

vegetarian variation:

Omit the smoked turkey.

2 large carrots sliced (about 1 cup)

1 cup green beans, cut into 1-inch pieces, steamed until tender, and chilled

1 cup sliced mushrooms (about 3 ounces)

½ cup sliced radishes

2 scallions, finely chopped (about ¼ cup)

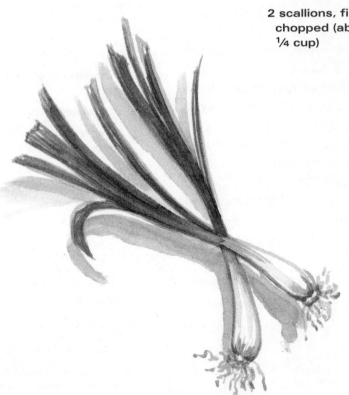

gazpacho bulgur salad

A refreshing salad—also good made with barley.

6 to 8 servings

1. Heat the water and $1/2$ cup tomato juice to almost boiling. Pour over the bulgur in a large bowl and let stand, covered, for 15 minutes. Fluff with a fork.

2. Combine the remaining tomato juice, the olive oil, vinegar, garlic, cumin, salt, black pepper, and Tabasco sauce in a small bowl. Add to the bulgur along with the scallions, tomatoes, cucumber, and bell pepper. Toss well. Taste and adjust seasoning. Chill before serving on a bed of greens.

1 cup water

¾ cup tomato juice

¾ cup bulgur, medium granulation

2 tablespoons olive oil

2 tablespoons balsamic or red wine vinegar

1 garlic clove, minced

½ teaspoon ground cumin

½ teaspoon salt

Freshly ground black pepper to taste

Few drops Tabasco or other hot red-pepper sauce

3 scallions, minced (about ⅓ cup)

2 ripe tomatoes, chopped or peeled, seeded, and chopped (about 2 cups)

1 cucumber, peeled, seeded, and diced (about 1 cup)

1 green bell pepper, cored, seeded, and diced (about ¾ cup)

Salad greens

warm wheat berry and lentil salad

4 servings

1. Bring 3 cups water to boiling in a 2-quart saucepan. Sort and rinse the wheat berries; add to the boiling water. Return the mixture to a boil; reduce the heat to low and cook, covered, 35 minutes.

2. Sort and rinse the lentils. When the wheat berries have cooked 35 minutes, stir in the lentils and cook 25 to 35 minutes longer or until wheat berries and lentils are as tender as you like them, adding additional water, if necessary.

3. Stir the carrot and zucchini into the wheat berry mixture. Set aside, covered, 5 minutes. Drain well.

4. Stir the vinegar, olive oil, sugar, and salt into the wheat berry mixture and serve warm.

3 to 4 cups water

½ cup wheat berries

½ cup green or brown lentils

½ cup coarsely shredded carrot (1 medium carrot)

½ cup coarsely shredded zucchini (½ small zucchini)

2 tablespoons cider vinegar

2 tablespoons olive oil

1 teaspoon sugar

½ teaspoon salt

pilafs

There are no fields of amaranth on this side of the grave.

—WALTER SAVAGE LANDOR, *Aesop and Rhodope*, 1846

chicken with sun-dried tomatoes, rice, and bulgur

Bonnie's friend Jacqueline Sirlin shared a multistep, multipot flavorful recipe with us. We used her flavor profile, simplified the instructions, and ended up with this unusual dish.

6 to 8 servings

1. Combine 1 tablespoon olive oil, the garlic, 1 teaspoon *each* cumin and coriander, the salt, and pepper in a bowl or plastic bag. Add chicken and let marinate for 45 minutes at room temperature or overnight in the refrigerator.

2. Cover the sun-dried tomatoes with boiling water and let stand until softened, about 10 minutes. Drain the tomatoes, reserving the liquid, and slice into slivers.

3. Heat remaining oil in 2-quart Dutch oven over medium-high heat. Add the onion and mushrooms and cook until the mushrooms give off their liquid and it evaporates; push

2 tablespoons olive oil

2 garlic cloves, minced

4 teaspoons ground cumin

4 teaspoons ground coriander

Salt and freshly ground black pepper, to taste

1½ pounds skinned and boned chicken breasts

1 ounce sun-dried tomatoes (about 10 to 12)

¼ cup boiling water

1 large onion, chopped (about 1½ cups)

½ pound mushrooms, sliced (about 3 cups)

½ cup bulgur, medium granulation

1 teaspoon minced fresh ginger

1 cup long grain white rice

3 cups chicken broth

1 teaspoon dried thyme

½ cup shelled pistachios

3 tablespoons chopped fresh cilantro

vegetables to the side of the pot. Add the bulgur and ginger; cook 1 minute. Add the rice and continue to cook 2 minutes, stirring constantly.

4. Preheat oven to 325° F. Pour in the broth, tomato liquid, tomato slivers, remaining cumin and coriander, the thyme, and salt and pepper to taste; mix well. Bring to a boil and remove from the heat. Remove chicken from bowl or bag, discarding the marinade, and place in the Dutch oven. Cover and bake for 20 minutes, until the chicken is cooked through and all liquid is absorbed. Remove from the oven and let stand 5 minutes. Taste and adjust seasoning.

5. Fluff grains with a fork, mound onto a platter, and top with chicken. Sprinkle with pistachios and cilantro.

bulgur and vermicelli pilaf

6 servings

1. Heat the oil in a medium saucepan over medium-high heat. Add the shallots and bell pepper; sauté until shallots soften, about 5 minutes. Add vermicelli and sauté until lightly colored, about 3 minutes. Add the bulgur and stir until toasted, 1 minute more.

2. Add the broth, tarragon, salt, and black pepper. Bring to a boil, reduce heat, and simmer 10 to 15 minutes, until liquid is absorbed and bulgur is tender. Taste and adjust seasoning. Fluff with a fork, sprinkle with parsley, and serve.

2 tablespoons olive oil

½ cup minced shallots

1 medium red bell pepper, seeded, cored, and chopped (about ¾ cup)

3 ounces vermicelli, broken into small pieces

1 cup bulgur, medium granulation

2 cups chicken or vegetable broth

1 teaspoon dried tarragon

Salt and freshly ground black pepper to taste

¼ cup chopped fresh parsley

irish oatmeal pilaf

4 servings

1. Heat oats in a large, dry skillet, stirring constantly over medium heat, 3 minutes or until very lightly toasted.

2. Add the oil, then stir in the bell pepper, carrot, scallions, mushrooms, and garlic and sauté, stirring constantly, 5 minutes or until the garlic begins to brown.

3. Stir in the curry powder, cumin, and salt until uniformly combined. Carefully add the water, stirring to combine. Bring the mixture to a boil; reduce heat to low, cover, and cook 20 minutes. Stir in the peas and cook 5 minutes longer or until the oats are tender. Turn the pilaf out onto a serving platter. Top with chopped parsley.

1 cup steel-cut (Irish) oats

1 tablespoon olive oil

½ medium red bell pepper, cored, seeded, and chopped (about ⅓ cup)

1 medium carrot, thinly sliced (about ½ cup)

2 scallions, thinly sliced (about ¼ cup)

4 medium mushrooms, thinly sliced (about ⅓ cup)

2 cloves garlic, sliced

1 teaspoon curry powder

½ teaspoon ground cumin

½ teaspoon salt

3 cups water

½ cup frozen peas

1 tablespoon chopped fresh parsley

green kasha pilaf

A delightful hearty accompaniment to any roast meat, poultry, or fish.

6 servings

1. Combine the egg and kasha in a small bowl until the grains are evenly coated. Heat a 2-quart saucepan over high heat, reduce heat to low, add the kasha and egg mixture, and stir constantly to break up any clumps and until the grains are roasted, 5 to 10 minutes. Remove from saucepan and set aside.

2. Heat the butter and oil in the saucepan over medium-high heat. Add the scallions and bell pepper and sauté until softened, about 5 minutes. Stir in the kasha, broth, salt, and black pepper. Bring to a boil over high heat, reduce heat to low, cover, and simmer 7 minutes.

3. Meanwhile, chop the spinach, parsley, and basil using a knife or food processor fitted with a steel blade. Stir into the kasha mixture. Cover and continue to cook until all the liquid is absorbed, about 5 minutes more. Taste and adjust seasoning. Sprinkle with Parmesan, if desired, and serve.

1 egg (or egg white), lightly beaten

1 cup roasted buckwheat groats (kasha), medium granulation

1 tablespoon butter

1 tablespoon olive oil

2/3 cup finely minced scallions, including green tops (about 4 scallions)

1 medium green bell pepper, cored, seeded, and chopped (about 3/4 cup)

2 cups chicken or vegetable broth

Salt and freshly ground black pepper

1 cup loosely packed spinach leaves

1/2 cup loosely packed parsley leaves

1/2 cup loosely packed basil leaves

1/4 cup freshly grated Parmesan cheese (optional)

quinoa and wild and brown rice pilaf

6 to 8 servings

1. Heat the butter in a 2-quart saucepan over medium heat. Add onion and celery; sauté until softened, about 5 minutes. Add brown rice; cook, stirring constantly, until the rice is coated with the butter, about 3 minutes.

2. Add the broth, sherry, tarragon, and soy sauce, then bring to a boil over high heat. Stir in the wild rice, cover, reduce heat to low, and cook 35 minutes.

3. While the rice is cooking, toast the quinoa over medium-high heat in a dry, small skillet, stirring constantly, until the grains darken and are aromatic. Quickly stir the quinoa into the rice mixture, cover, and simmer for 15 minutes, until the quinoa and rice are tender and the liquid is absorbed. Remove from the heat and allow to stand covered for 5 minutes. Add parsley, salt, and pepper. Fluff with a fork. Taste and adjust seasoning and serve.

1 tablespoon butter

1 medium onion, chopped (about 1 cup)

1 rib celery, chopped (about ⅔ cup)

½ cup long grain brown rice

3½ cups vegetable broth or water

¼ cup dry sherry

1 tablespoon fresh or 1 teaspoon dried tarragon

1 teaspoon soy sauce

½ cup wild rice, picked over, rinsed in cold water, and drained

½ cup quinoa, rinsed in cold water and drained

¼ cup chopped fresh parsley

Salt and freshly ground black pepper to taste

mixed-grain pilaf
with fruit

This is an especially good choice to serve with duck, goose, or game.

6 servings

1. Bring 2 cups water to a boil in a 1-quart saucepan. Heat 1 tablespoon of the amaranth over high heat in a large, dry skillet with a lid, shaking until the grain pops. Turn popped amaranth out into a small bowl and set aside. Reserve the skillet.

2. Add the remaining amaranth and the teff to the boiling water. Return the mixture to a boil; reduce heat to low and cook, covered, 10 minutes.

3. Meanwhile, heat the oil over medium heat in the large skillet. Add the millet, rice, and onion to the skillet and sauté, stirring constantly, 3 minutes, or until the grains start to brown.

4. Add the amaranth mixture, along with any water remaining in the saucepan, and the salt and remaining 2 cups water to the mixture in the skillet. Bring the mixture to a boil over high heat; reduce heat to low and cook, covered, 10 minutes.

4 cups water

¼ cup amaranth

¼ cup teff

1 tablespoon vegetable oil

½ cup millet

½ cup basmati or other aromatic rice

1 small onion, chopped (about ½ cup)

¾ teaspoon salt

¼ cup dried sour cherries or cranberries

¼ cup coarsely chopped dried peaches or apricots

5. Stir the cherries and peaches into the pilaf mixture in the skillet. Cook 5 to 10 minutes longer, stirring constantly, until the grains have reached the desired tenderness and the water has been absorbed.

6. Turn the pilaf out into a serving bowl and top with the reserved popped amaranth.

mixed-grain and wild mushroom pilaf

The earthiness of mushrooms marries so well to grains, we decided to combine several. The results were delicious.

8 servings

1. Heat the oil in 3-quart saucepan over medium-high heat. Add the shallots and mushrooms and cook until the mushrooms give off their liquid and it evaporates. Add the amaranth, millet, rice, and bulgur. Stir to coat lightly with the oil and cook 1 minute, until browned lightly.

2. Add the broth, salt, and pepper. Bring to a boil over high heat, reduce heat to low, cover, and let cook about 15 minutes, until the liquid is absorbed and the grains are tender. Let stand 5 minutes, then fluff with a fork and sprinkle with parsley. Taste and adjust seasoning. Sprinkle with cheese and serve.

1 tablespoon olive oil

½ cup minced shallots

12 ounces cultivated exotic (wild) mushrooms, such as chanterelles, morels, porcini, portobellos, and/or shiitakes, sliced

2 tablespoons amaranth

¼ cup millet

¼ cup long grain white rice

¼ cup bulgur, medium granulation

2½ cups chicken, beef, or vegetable broth

Salt and freshly ground black pepper to taste

¼ cup chopped fresh parsley

Freshly grated Parmesan cheese

mushroom barley pilaf

A delicious side dish to accompany any roast.

6 servings

1. Heat the oil in a 3-quart saucepan over medium-high heat. Add the onion and mushrooms and cook until the mushrooms give off their liquid and it evaporates. Add the barley, stir to coat lightly with the oil, and cook 1 minute, until browned lightly.

2. Add the broth, salt, and pepper. Bring to a boil, reduce heat, cover, and cook 35 to 45 minutes, until the liquid is absorbed and the barley is tender. Let stand 5 minutes, then fluff with a fork. Taste and adjust seasoning and serve.

2 tablespoons olive oil

1 large onion, chopped (about 1½ cups)

1 pound mushrooms, sliced (about 6 cups)

1 cup pearl barley, rinsed and drained

2 ¾ cups chicken or vegetable broth

Salt and freshly ground pepper to taste

side dishes

How like, methought,

I then was this

kernel,

This squash, this

gentleman

—WILLIAM

SHAKESPEARE,

The Winter's Tale,

1610–11

barley provençale

The taste of France in a new form.

6 servings

1. Heat the oil in a 2-quart saucepan over medium-high heat. Add the onion and cook until translucent, about 5 minutes. Add the garlic and barley and continue to cook, stirring, until barley is lightly browned, about 2 minutes.

2. Add the broth, tomato, half the parsley, the capers, salt, and pepper. Bring to a boil over high heat, cover, reduce heat to low, and cook until liquid is absorbed, 35 to 40 minutes. Add the remaining parsley. Taste and adjust seasoning and serve.

1 tablespoon olive oil

½ large onion, chopped (about ¾ cup)

2 garlic cloves, minced

¾ cup pearl barley, rinsed and drained

1¼ cups chicken or vegetable broth

1 large ripe tomato, peeled and diced, or 1 cup canned crushed tomatoes

¼ cup chopped fresh parsley

1 tablespoon capers, rinsed and drained

Salt and freshly ground black pepper to taste

barley with fennel

For fennel lovers everywhere! Serve with broiled or grilled fish, chicken, or steak.

4 to 6 servings

2 tablespoons olive oil

1 fennel bulb, trimmed and finely chopped (about 3 cups)

½ cup minced shallots

2 garlic cloves, minced

1 cup pearl barley, rinsed and drained

1¾ cups chicken or vegetable broth

Salt and freshly ground black pepper

½ cup freshly grated Parmesan cheese

2 tablespoons chopped fresh parsley

1. Heat the oil in a 4-quart saucepan over medium-high heat. Add the fennel, shallots, and garlic; reduce heat to low, and cook gently until fennel is tender, about 5 minutes.

2. Stir in the barley and cook, while stirring, 1 minute more. Add the broth, salt, and pepper; bring to a boil, reduce heat to low, cover, and cook 45 minutes or until barley is tender and water is absorbed.

3. Remove from heat, sprinkle with cheese and parsley, and let stand, covered, 2 minutes before serving.

grits and grains soufflé

4 servings

1. Preheat the oven to 375° F. Lightly grease a shallow 1-quart casserole.

2. Separate the eggs; place the whites in a small bowl and yolks in a large bowl.

3. With an electric mixer, beat the egg whites until stiff peaks form.

4. With same beaters, gradually beat the grits, wheat cereal, teff, and milk into the egg yolks. Reserve 2 tablespoons cheese. Fold remaining cheese, the pepper, and one-third of the beaten egg whites into the grain mixture until uniformly combined.

5. Carefully fold in the remaining beaten whites just until combined. Turn the mixture into the prepared casserole. Sprinkle reserved cheese over top.

6. Bake soufflé 45 to 50 minutes or until well browned and center appears set when the casserole is gently tapped. Serve immediately

note: *Cook grits, cereal, and teff in salted water according to package directions.*

2 eggs

1 cup cooked natural corn grits (see note)

¾ cup cooked cracked wheat cereal (see note)

¼ cup cooked teff or additional cracked wheat cereal, warm (see note)

½ cup milk

½ cup shredded Cheddar cheese

¼ teaspoon cracked black pepper

great grits

Good any time of the day, grits are the ultimate convenience food.
You can cook them quickly or cook them ahead and reheat them. You
can eat them for breakfast, lunch, or dinner; serve them as a soufflé
for company; or cook yourself a bowl in the middle of the night when
you need a hug from home. As long as you can boil water, you can
cook grits.

Although the dictionary defines grits as "ground hominy with the
germ removed," the lexicographers obviously haven't noticed the wide
variety of grits on the market today. The most readily available com-
mercial grits are now made from ground white or yellow corn with
the germ removed. However, they are not usually made from hominy.
In addition to rye, barley, rice, and kamut grits, health food stores
and mail order catalogs sell delicious whole corn grits, which don't
have the germ removed.

kasha varnishkes

A grain book wouldn't be complete without this traditional classic European Jewish dish. A perfect accompaniment for braised beef.

6 to 8 servings

1 egg (or egg white), lightly beaten

1 cup roasted buckwheat groats (kasha), medium granulation

1 tablespoon oil

1 large onion, chopped (about 1½ cups)

2 cups chicken or vegetable broth

Salt and freshly ground black pepper to taste

2 cups bow-tie pasta, cooked al dente

2 tablespoons butter

1. Combine the egg and kasha in a small bowl until the grains are evenly coated. Heat a 2-quart saucepan over high heat, reduce heat to low, add the kasha and egg mixture, and stir constantly to break up any clumps and until the grains are roasted, 5 to 10 minutes. Transfer kasha to a bowl and set aside.
2. Heat the oil in the same saucepan over high heat. Add the onion, reduce the heat to medium, and cook until browned, 10 to 15 minutes. Transfer to a bowl and set aside.
3. Add the kasha, broth, salt, and pepper to the saucepan; bring to a boil over high heat, reduce heat to low, and cook until the broth is absorbed, 10 to 15 minutes.
4. Toss the pasta with the butter, onions, and kasha. Taste and adjust seasoning. Serve.

three-grain hush puppies

6 servings

1. Heat enough oil in a 1-quart saucepan or a deep fryer to fill about 3 inches.

2. Combine the flour, cornmeal, rolled oats, scallion, baking powder, and salt in a medium bowl. Stir in the milk, egg, and Tabasco sauce with a fork until a soft dough forms.

3. When the oil reaches 375° F. on a deep fat thermometer, drop the dough by slightly rounded teaspoonfuls, several at a time, into the fat and fry until golden brown on all sides. Drain on paper towels and serve.

Vegetable oil for frying

½ cup spelt or whole wheat flour

½ cup cornmeal

½ cup rolled oats

1 scallion, finely chopped (about 2 tablespoons)

2 teaspoons baking powder

¼ teaspoon salt

½ cup milk

1 egg

2 to 4 drops Tabasco or other hot red-pepper sauce

parmesan gnocchi

The character of this dish is determined by the cheese you use. So, use the very best you can find.

4 servings

1. Separate the eggs; place the whites in a small bowl and the yolks in a large bowl.

2. Combine the milk, semolina, and butter in a 1-quart saucepan. Bring the mixture to a boil, stirring with a fork. Cook, stirring constantly, until thickened, 2 minutes.

3. Meanwhile, bring about 3 inches of salted water to a boil in a 3-quart saucepan.

4. Gradually beat the semolina mixture, $^1/_4$ cup of the cheese, and the chopped basil into the yolks. With an electric mixer, beat the egg whites until soft peaks form. Fold the beaten egg whites into the semolina mixture.

5. To make gnocchi, fill a measuring teaspoon with semolina mixture so it is slightly rounded. Hold the spoon about an inch above the boiling water; with your finger gently push the mixture into the water, being careful not to cause the water

2 eggs

1¼ cups milk

½ cup coarse semolina

2 tablespoons butter

½ cup coarsely grated Parmesan cheese

½ tablespoon chopped fresh or ½ teaspoon dried basil

4 cloves garlic sliced

3 tablespoons olive oil

1 tablespoon thinly sliced fresh basil leaves (optional)

to splash. Repeat to use about one-fourth of the semolina mixture. Cook gnocchi until they rise to the surface; then cook about 1 minute longer. Remove with a slotted spoon and let drain in a colander. Repeat until all gnocchi have been cooked.

6. Preheat oven to 350° F. Sauté the garlic in the oil in a small skillet over medium heat, until it begins to brown; remove from heat. Stir in the sliced basil, if desired. When all gnocchi have been cooked, place them in a shallow 1-quart casserole. Top with the garlic mixture and the remaining $1/4$ cup of cheese. Bake 20 minutes or until gnocchi have been reheated through and the cheese has melted. Serve.

what's in a name?

While Parmigiano-Reggiano is Parmesan cheese, not all Parmesan cheese is Parmigiano-Reggiano. Made from a seven-hundred-year-old formula, this distinctive cheese comes only from prescribed areas near Parma and Reggio Emilia in Italy. Because it is made from partially skimmed milk, Parmigiano-Reggiano is the lowest in fat and cholesterol of any aged cheese. Because of its special handling, it is completely natural, preservative free, and sublimely delicious.

squash and barley casserole

A hearty home-style casserole that's unusually delectable with a simple roast.

6 to 8 servings

1. Heat the oven to 350° F. Heat the oil in a 4-quart saucepan or oven-proof casserole over medium-high heat. Add the onion, bell pepper, and garlic and cook over medium heat until the onions are translucent, about 5 minutes.

2. Stir in the barley and cook, stirring, 1 minute more. Add the broth, squash, currants, zest, salt, allspice, cinnamon, cayenne, and black pepper; bring to a boil. Cover, transfer to oven, and bake 40 to 45 minutes, until barley is tender and water is absorbed, stirring once during cooking.

3. Remove from oven, taste and adjust seasoning, fluff with a fork, cover, and let stand 5 minutes. Sprinkle with almonds and serve.

1 tablespoon vegetable oil

1 large onion, chopped (about 1½ cups)

1 medium green bell pepper, cored, seeded, and finely chopped (about ¾ cup)

1 garlic clove, minced

1 cup pearl barley, rinsed and drained

2 ¾ cups chicken or vegetable broth

1 butternut squash, peeled, seeded, and diced (about 3 cups)

¼ cup currants

2 teaspoons grated orange zest

1 teaspoon salt

½ teaspoon ground allspice

½ teaspoon ground cinnamon

Pinch cayenne (ground red) pepper

Freshly ground black pepper, to taste

½ cup almonds, toasted and chopped

main dishes

The Farmer will

never be happy

again;

He carries his heart

in his boots;

For either the rain is

destroying his grain

Or the drought is

destroying his roots.

—Sir Alan Patrick

Herbert, *The*

Farmer, 1922

chicken posole

This aromatic stew appears in our cover photo.

6 servings

1. Sauté the chicken in the oil in a large skillet over medium heat until lightly browned, about 5 minutes.

2. Add the onion, bell peppers, and garlic to the pan. Cook, stirring frequently, until the vegetables are just tender and the chicken is cooked through, 15 to 20 minutes.

3. Stir the posole and chilies into the vegetable mixture along with the basil, thyme, Tabasco sauce, salt, and black pepper. Cook, stirring constantly, until the posole is heated through. Taste and adjust seasoning. Top with cilantro, if desired, and serve immediately.

2 whole chicken breasts, skinned and boned and cut into 1-inch pieces (about 1 pound)

1 tablespoon olive oil

1 medium onion, chopped (about 1 cup)

1 medium green bell pepper, cored, seeded, and chopped (about ¾ cup)

1 medium red bell pepper, cored, seeded, and chopped (about ¾ cup)

3 garlic cloves, sliced

2 cups cooked posole (see box page 83) or well-drained canned hominy

½ 4-ounce can chopped mild green chilies

1 teaspoon dried basil

¼ teaspoon dried thyme

4 to 6 drops Tabasco sauce or other hot-pepper sauce

Salt and freshly ground black pepper to taste

2 tablespoons chopped fresh cilantro (optional)

hominy or posole

To cook hominy or posole, rinse and remove any foreign materials from 1 cup. Combine with 5 cups water and 1 teaspoon salt; cover and refrigerate overnight. The next day, place hominy and soaking water into a 3-quart saucepan. Bring to a boil over high heat; reduce heat, cover, and simmer 5 to 6 hours or until tender, adding water as necessary. Or, combine 1 cup rinsed and cleaned hominy or posole, 5 cups boiling water, and 1 teaspoon salt in a 2- or 3-quart Crock-Pot. Cover and cook on low heat 6 to 8 hours.

buckwheat noodles in spicy peanut sauce

6 servings

1. Bring salted water to a boil in a large saucepan for the noodles.

2. Meanwhile, heat the oil in a large skillet over medium heat. Add the carrot, scallions, and garlic; sauté 3 to 5 minutes or until carrots are just tender. Stir in the peanut butter, boiling water, soy sauce, sugar, and Tabasco sauce. Remove from heat and set aside. Peel and quarter cucumber. Remove and discard seeds; cut cucumber into thin julienne strips.

3. Cook the noodles in the boiling water according to package directions. Be careful not to overcook. Drain well and toss immediately with the peanut sauce. Turn out onto a serving platter and top with cucumbers and cilantro. Serve hot or refrigerate, covered, and serve cold.

1 tablespoon Oriental sesame oil or peanut oil

1 large carrot, cut into very thin julienne strips

2 scallions, cut into very thin julienne strips

4 garlic cloves, sliced

⅓ cup chunky peanut butter

1 cup boiling water

1 tablespoon soy sauce

1 teaspoon sugar

2 to 4 drops Tabasco sauce or other hot red-pepper sauce

½ small cucumber

1 (8- to 9-ounce) package 100% buckwheat (soba) noodles

1 tablespoon chopped fresh cilantro

buckwheat, or soba, noodles

Buckwheat, or soba, noodles may be found in Oriental markets and health food stores. Although they usually contain buckwheat, soba noodles may also contain wheat flour, potato starch or cornstarch, green tea, eggs, yams or mushrooms.

aromatic chicken and quinoa

4 servings

1. Heat the oil in a medium-large, deep skillet over medium-high heat. Add the shallots, celery, and carrots and sauté until shallots are translucent, about 5 minutes. Stir in the garlic and parsley; cook 1 minute more. Push the vegetables to the side of the skillet. Increase heat to high.

2. Season the chicken with thyme, salt, and pepper. Add to the skillet and cook, stirring constantly, until lightly colored. Add the broth and sherry, bring to a boil over high heat, reduce heat to low, cover and simmer 10 minutes. *This recipe can be made ahead until this point. Refrigerate until ready to continue.*

3. If refrigerated, return to skillet and bring to a simmer.

4. Stir in quinoa, cook 15 minutes, remove from heat, and let stand 5 minutes. Serve.

1 tablespoon olive oil

1 ½ cups chopped shallots

2 ribs celery, chopped (about 1 ⅓ cups)

3 carrots, chopped (about 1 ½ cups)

2 garlic cloves, minced

½ cup minced fresh parsley

1 pound skined and boned chicken breasts, cut into 1-inch chunks

1 teaspoon dried or 1 tablespoon fresh thyme

Salt and freshly ground black pepper to taste

2 cups chicken broth

½ cup dry sherry

¾ cup quinoa, rinsed in cold water and drained

red pepper polenta with ratatouille

4 servings

1. Heat the sausage in a heavy-bottomed saucepan over medium heat until cooked through. Remove and set aside. Discard all but 1 teaspoon fat from the pan. Add the onion and cook over medium heat, stirring, until soft and pale golden, about 5 minutes. Add the garlic, eggplant, tomatoes, and tomato puree. Cover and cook, stirring occasionally, for 10 to 15 minutes, until eggplant is tender.

2. Stack the basil leaves. Roll them up and thinly slice with a sharp knife. Stir half the basil and half the parsley into the saucepan along with the zucchini, yellow squash, salt, and pepper and cook an additional 10 minutes. Serve the ratatouille over the Red Pepper Polenta, garnished with the remaining parsley and basil, and the reserved sausage.

2 ounces andouille sausage, sliced paper thin

1 medium onion, chopped (about 1 cup)

1 garlic clove, minced

½ pound white eggplant, peeled and diced (about 2 cups)

2 large ripe tomatoes, peeled and diced (about 2 cups)

3 tablespoons tomato puree

¼ cup basil leaves, loosely packed

2 tablespoons chopped fresh parsley

1 small zucchini, finely diced (about 1 cup)

1 yellow squash, finely diced (about 1 cup)

Coarse salt and freshly ground pepper to taste

Red Pepper Polenta (recipe follows)

red pepper polenta

4 servings

1. Whisk together the cornmeal, water, salt, and crushed red pepper in a medium bowl. Bring the broth to a boil in a medium pot. Add the cornmeal mixture in a slow stream, stirring constantly with a whisk. Reduce heat to low, and using a wooden spoon, stir until mixture begins to thicken. Simmer over medium heat for 10 to 15 minutes, stirring frequently, until the consistency of cooked cereal. Stir in the cheese, cover, remove from heat, and let stand 5 minutes.

1 cup cornmeal

1 cup cold water

¼ teaspoon coarse salt

¼ teaspoon crushed red pepper

1½ cups vegetable broth

2 tablespoons freshly grated Parmesan cheese

couscous and chicken madeira

An easy-to-prepare dish, great for entertaining.

6 to 8 servings

1. Heat the oil over high heat in a 4-quart saucepan, add the mushrooms and cook, stirring constantly, about 2 minutes. Add the onions and continue to cook, stirring constantly, about 2 minutes. Sprinkle with the sugar, reduce heat to medium, and cook, stirring occasionally, until the mushrooms and onions are glazed, about 5 minutes.

2. Place the chicken over the mushrooms and onions, add the broth, wine, tarragon, and garlic. Bring to a boil, reduce the heat to low, cover, and let simmer for 40 to 45 minutes, until chicken is cooked through. Taste and adjust seasoning. Pour off $2^{1}/4$ cups cooking liquid into a measuring cup.

1 tablespoon olive oil

1 (10-ounce) package button mushrooms

1 pound fresh or frozen small white onions (see note)

2 tablespoons packed dark brown sugar

2 pounds chicken thighs, skinned

1 cup chicken broth

½ cup Madeira wine

1 teaspoon dried tarragon

2 garlic cloves, minced

1½ cups couscous

½ cup finely minced fresh parsley

This recipe can be made ahead until this point. Refrigerate until ready to continue.

3. If refrigerated, bring cooking liquid to a boil. Reheat chicken mixture.

4. Put couscous into a large bowl. Pour cooking liquid over top, cover tightly and let stand 5 to 10 minutes, until all the liquid is absorbed. Fluff with a fork. Mound on a serving platter, spoon chicken mixture around the couscous, sprinkle with parsley, and serve.

note: *If using fresh small white onions, place in boiling water for 1 minute, then quickly into cold water to cool. Trim the ends and the skins should slip right off.*

fresh corn empanadas

4 servings

1\. Combine the *masa harina*, $1^1/_2$ cups bread flour, the sugar, salt, and yeast in a large bowl. Stir the water and oil into the flour mixture until a soft smooth dough forms.

2\. Turn the dough out onto a lightly floured board and knead for 3 minutes, adding as much of the remaining bread flour as necessary to make the dough manageable. Shape the dough into a ball on the board and cover it with the bowl. Let rise until double in bulk, 20 to 25 minutes.

3\. When the dough has risen, grease a large baking sheet. Divide the dough into 4 balls; roll out each to make a 6-inch round; moisten the edges of the rounds. Set aside 2 tablespoons cheese. Combine the remaining cheese, the corn, and salsa in a medium bowl; divide onto the dough rounds. One at a time, fold the dough

$1^1/_4$ cups *masa harina* or corn flour

$1^1/_2$ to 2 cups bread or all-purpose flour

2 teaspoons sugar

1 teaspoon salt

1 package ($2^1/_4$ teaspoons) fast-rising dry yeast

1 cup hot water (120° to 125° F.)

1 tablespoon vegetable oil

1 cup shredded Monterey Jack or jalapeño Monterey Jack cheese

1 cup fresh corn kernels

$^1/_2$ cup prepared thick salsa (see note)

rounds in half over the corn mixture to make empanadas; pinch the moistened edges together to seal.

4. Place the empanadas on the greased baking sheet. Cover them lightly with a linen towel and set aside in a warm place to rise until double in bulk, 25 to 30 minutes.

5. Preheat the oven to 375° F. Sprinkle the reserved cheese over the empanadas. Bake the empanadas until they sound hollow when gently tapped on the top, 25 to 30 minutes. Serve.

note: *Prepare home-made salsa or use your favorite commercial one. If you cannot find a thick one, strain off most of the liquid so the empanada filling is not too juicy.*

hominy and sausage

6 servings

1. Pierce the sausages several times with a fork. Bring the sausages and water to cover to a boil in a large skillet over high heat. Reduce heat to medium, cover, and cook sausages 15 minutes. Drain.

2. Heat the sausages, uncovered, in skillet over low heat until skillet is dry, then add the butter, onion, and bell pepper. Cook, stirring vegetables and turning sausages until vegetables are just tender and sausages are browned on all sides, about 5 minutes.

3. Stir the hominy into the vegetable mixture along with salt and black pepper and cook, stirring constantly, until hominy is heated through. Taste and adjust seasoning. Serve immediately.

6 links fresh pork sausage (about 1½ pounds)

1 teaspoon butter

1 medium onion, chopped (about 1 cup)

1 medium green bell pepper, cored, seeded, and chopped (about ¾ cup)

3 cups cooked hominy (see pages 16, 18) or well-drained canned hominy

Salt and freshly ground black pepper to taste

irish ris-oat-o

Steel-cut, or Irish, oats cook to the same creaminess as arborio rice.

4 servings

1. Melt the butter in a 2-quart saucepan over medium heat. Add the onion and garlic and sauté until soft, about 5 minutes; add the oats and cook over medium heat about 3 minutes until lightly toasted.
2. Add the hot broth 1 cup at a time, stirring constantly with a wooden spoon and waiting until the broth is absorbed before adding more. Continue adding broth 1 cup at a time and stirring until 6 cups have been added, about 20 minutes in all. Stir in the carrot and scallions and cook 5 minutes longer or until the oats are cooked but still al dente, or firm to the bite. Stir in some of the remaining broth as needed, to keep the mixture moist and creamy.
3. Remove from heat; stir in the cheese. Taste and add salt if necessary; serve immediately.

2 tablespoons butter

1 small onion, finely chopped (about ½ cup)

1 garlic clove, finely chopped

1½ cups steel-cut (Irish) oats

6 to 7 cups chicken or vegetable broth, simmering

1 medium carrot, shredded

2 scallions, sliced crosswise

½ cup grated white Cheddar cheese

Salt (optional)

kamut tagliatelle with roasted vegetables

This is a softer-than-usual pasta dough, so it can be easily rolled out by hand to make a tender, wheaty pasta.

4 servings

1. Combine 1½ cups kamut flour, ½ teaspoon basil, and ½ teaspoon salt in a large bowl. Beat the egg, water, and 2 tablespoons oil in a measuring cup. Stir the water mixture into the flour mixture with a fork until a soft, smooth dough forms.
2. Turn the dough onto a lightly floured board and knead for 5 minutes, adding only as much of the remaining flour as necessary to keep the dough from being sticky. Shape the dough into a ball on the board and cover it with the bowl. Let the dough rest 30 minutes.
3. Meanwhile, make the sauce. Preheat oven to 400° F. Toss the zucchini, yellow squash, bell pepper, onion, mushrooms, garlic, 1 tablespoon of the remaining oil, the remaining ¼ teaspoon basil and ¼

1½ to 2 cups kamut or semolina flour

¾ teaspoon dried basil

¾ teaspoon salt

1 egg

⅓ cup cold water

¼ cup olive oil

1 small (5-inch) zucchini, cut into ½-inch chunks (about 1¼ cups)

1 small (5-inch) yellow squash, cut into ½-inch chunks (about 1¼ cups)

1 small red bell pepper, coarsely chopped (about ½ cup)

1 small onion, chopped (about ½ cup)

6 medium (¼-pound) mushrooms, sliced (about 1½ cups)

4 cloves garlic, sliced

¼ teaspoon cracked black pepper

½ cup shredded Parmesan cheese

teaspoon salt, and the black pepper in a large bowl until combined. Arrange the vegetables on 2 rimmed baking sheets. Roast 15 to 20 minutes or until vegetables are tender.

4. Bring 4 quarts salted water to a boil in a 6-quart saucepan. Divide the dough in half. Roll out each half to form a 12-inch square. Fold the square in half; roll out to make a 12-inch square again. Fold in half in the other direction and roll out to form a 14-inch square. Generously flour both sides of the dough. Slice into tagliatelle strips, 14 inches long and $1/4$ to $1/2$ inch wide.

5. Cook tagliatelle until just tender, about 5 minutes. Drain very well. Toss with the remaining 1 tablespoon oil and divide onto serving plates. Top with the roasted vegetables and cheese.

mixed-grain and vegetable pie

6 servings

1. Combine the flour, oats, $^1/_2$ teaspoon salt, and the black pepper in a large bowl. With a fork stir in 3 tablespoons oil until dry ingredients are moistened. Turn the mixture into a 9-inch pie plate and pat to form an even layer on the bottom and up the sides of the plate. Reserve the bowl.

2. Preheat the oven to 350° F. Heat the remaining 1 tablespoon oil in a large skillet over medium heat. Add the onion, bell pepper, zucchini, carrot, and mushrooms to the skillet and sauté, stirring constantly, 5 minutes or until the onion begins to brown.

3. Combine the barley, millet, wheat berries, eggs, yogurt, remaining $^1/_2$ teaspoon salt, the basil, and thyme in the large bowl. Fold in the vegetables. Turn the mixture into the prepared crust. Set the pie on a rimmed baking sheet and bake 40 to 45 minutes or until the filling has set. Remove from the oven; cut into wedges and serve immediately.

1 cup whole wheat flour

⅓ cup rolled oats

1 teaspoon salt

¼ teaspoon cracked black pepper

¼ cup vegetable oil

1 small onion, chopped (about ½ cup)

1 small red bell pepper, cored, seeded, and chopped (about ½ cup)

½ medium zucchini, chopped (about ½ cup)

1 medium carrot, shredded (about ½ cup)

4 medium mushrooms, chopped (about ⅓ cup)

½ cup cooked barley (see pages 16, 18)

½ cup cooked millet (see pages 16, 18)

½ cup cooked wheat berries (see pages 16, 18)

2 eggs

1 (8-ounce) container plain nonfat yogurt

½ teaspoon dried basil

¼ teaspoon dried thyme

lamb and oats stew

4 servings

1. Sauté the lamb in olive oil in a heavy 4-quart Dutch oven or saucepan over medium heat until browned on all sides. Add the onion, garlic, salt, thyme, and pepper. Sauté, stirring constantly, for an additional 2 to 3 minutes. Carefully add 2 cups water. Bring the mixture to a boil; reduce the heat to low, cover the saucepan, and simmer the meat for 35 minutes.

2. Meanwhile, in a heavy ungreased skillet, toast the oats until lightly browned, about 3 minutes, stirring constantly. Add the mushrooms and butter to the skillet. Sauté, stirring, until mushrooms are lightly browned, about 5 minutes.

3. Stir the oat mixture into the simmering lamb mixture along with the carrots. Return the mixture to a boil. Cook 25 minutes longer, or until the lamb and oats are tender, adding more water if necessary. Serve from the Dutch oven or spoon onto a serving platter.

1 pound boneless lamb, cut into 1-inch pieces

1 teaspoon olive oil

1 medium onion, coarsely chopped (about 1 cup)

1 garlic clove, chopped

½ teaspoon salt

¼ teaspoon dried thyme leaves

¼ teaspoon cracked black pepper

2 to 3 cups water

1 cup steel-cut (Irish) oats

½ pound cultivated exotic (wild) mushrooms, such as shiitake or crimini, sliced (about 3 cups)

1 teaspoon butter

2 carrots, thinly sliced crosswise

moroccan lamb stew

4 to 6 servings

1. Heat 1 tablespoon oil in a heavy-bottom saucepan over medium-high heat. Brown the lamb, remove, and set aside. Heat the remaining oil in the pan. Add the onion and sauté until it is translucent, about 5 minutes. Add the garlic and cook, stirring, 1 minute more.

2. Mix in the carrots, tomatoes, coriander, saffron, and cayenne. Bring to a boil over high heat; reduce heat to low; stir in lamb. Cover and simmer for 25 minutes.

This recipe can be made ahead until this point. Refrigerate until ready to continue.

3. If refrigerated, bring mixture to a simmer.

4. Stir in the zucchini, yellow squash, and garbanzos (chick-peas) and simmer an additional 10 minutes. Taste and adjust seasoning.

5. In a medium bowl, combine couscous, currants, cinnamon, and tur-

2 tablespoons olive oil

1½ pounds boneless lean lamb, cut into 1-inch cubes

1 large onion, sliced (about 1½ cups

2 garlic cloves, minced

4 medium carrots sliced (about 2 cups)

2 cups canned crushed tomatoes

1 teaspoon ground coriander

½ teaspoon saffron threads

¼ teaspoon cayenne (ground red) pepper

1 medium zucchini, cut into chunks (about 1 cup)

1 medium yellow squash, cut into chunks (about 1 cup)

1 cup cooked or canned garbanzos (chick-peas), drained

1 cup couscous

¼ cup currents

meric. Pour broth over; stir and let stand, covered, 10 minutes or until liquid is absorbed and couscous is tender.

6. Mound couscous in the center of a serving dish, arrange the lamb and vegetables around it, and pour the stew liquid over all.

vegetarian variation:

Omit the lamb.

¼ teaspoon ground cinnamon

½ teaspoon ground turmeric

1½ cups boiling beef broth

---

classic couscous

Couscous is simply Moroccan pasta, which recently became popular in America. Like pasta, it's made with semolina flour and water. In Morocco, women sit on the ground with a large bowl and a sieve in their laps, working the dough together carefully with their hands until small pellets fall from their hands through the sieve and into the bowl. They cook the grains in the top section of a two-tiered pot called a couscousiere, *with the sauce cooking in the bottom. The vapors from the sauce steam and swell the grains on the top, perfuming them with the fragrance of the rich sauce. The grains are then spread out in a pan to dry, rubbed between the fingers to break up any clumps, then returned for another steaming. This process can be done up to seven times, although two or three are adequate.*

It's not necessary to own a couscousiere *to make this dish the authentic way. A good substitute is a colander fitted into a stockpot. Just be certain the simmering broth doesn't touch the bottom of the colander. To prevent steam from escaping, wrap a piece of cheesecloth or kitchen towel around the rim of the pot before inserting the colander.*

uncle herb's brisket with kasha and rice

Bonnie's uncle Herb is well known for his cooking—especially his delicious brisket. A great dish for entertaining, because you cook the meat a day before serving it.

8 to 10 servings

1. Make the brisket the day before you want to serve it. Preheat the oven to 500° F. Sprinkle the meat with the onion powder, garlic powder, paprika, salt, and pepper. Place into a deep baking pan and into the oven to brown, about 10 minutes.

2. Meanwhile, make sauce. Heat the tomatoes, 2 cups water, the soup mix, carrots, and celery in a 2-quart saucepan over medium-high heat and simmer until the vegetables are tender, about 10 minutes. Mix in the ketchup.

3. Remove the brisket from the oven and lower the heat to 350° F. Pour the sauce over the brisket, cover the pan tightly with foil, and bake $2^1/2$ to 3 hours, until the meat is tender. Pour off the sauce into a container

4-pound brisket of beef

Onion powder to taste

Garlic powder to taste

Ground paprika to taste

Salt and freshly ground black pepper to taste

2 cans (1-pound each) stewed tomatoes

6 cups water

1 (1 ounce) package onion soup mix

2 carrots, peeled and chunked (about 1 cup)

2 ribs celery, chopped (about 1 1/3 cups)

1/3 cup ketchup

1 cup roasted buckwheat groats (kasha)

1 cup long grain white rice

and place the brisket in a flat pan; refrigerate overnight.

4. The next day, remove and discard the hardened fat. Slice the brisket thinly across the grain; reheat in the sauce.

5. Bring the remaining 4 cups of water to a boil over high heat in a 2-quart saucepan. Stir in the buckwheat groats (kasha), rice, and salt; reduce heat to low, cover, and simmer 15 minutes, until the liquid is absorbed. Serve the brisket sauce over the kasha and rice along with the sliced meat.

wheat and turkey loaf

4 servings

1. Bring the tomatoes to a boil in a 1-quart saucepan. Stir in the bulgur and set aside 20 minutes.

2. Preheat the oven to 350° F. Lightly grease a 9 × 5-inch loaf pan.

3. Set aside 1 tablespoon onion. Combine the turkey, remaining onion, wheat berries, parsley, chili powder, salt, oregano, and pepper in a large bowl. Add the bulgur mixture and stir until completely combined.

4. Spoon the meat mixture into the prepared pan and spread to make even. Spread the ketchup over the top of the loaf and sprinkle it with the reserved onion.

5. Bake the meat loaf, uncovered, 55 to 60 minutes or until cooked through. Center should reach an internal temperature of 185° F. Remove to a serving platter and slice or refrigerate to serve cold.

1 (8-ounce) can stewed tomatoes

½ cup bulgur wheat, medium granulation

1 medium onion, chopped (about 1 cup)

1 pound ground turkey

½ cup cooked wheat berries

2 tablespoons chopped fresh parsley

2 teaspoons chili powder

1 teaspoon salt

½ teaspoon dried oregano

¼ teaspoon ground black pepper

3 tablespoons ketchup

breads

I keep picturing all these little kids play-ing some game in this big field of rye and all. . . . I mean if they're running and they don't look where they're going I have to come out from somewhere and catch them. That's all I'd do all day. I'd just be a catcher in the rye.

—J. D. SALINGER, *Catcher in the Rye,* 1951

anadama bread

There are many variations to the story of the fisherman damning his wife Anna for serving nothing but cornmeal mush and molasses and making his own bread with the two offending ingredients. So why was he at home baking bread instead of out fishing? Then they might have had fish to eat. Anyway, if the results were anything like this loaf, he was a pretty good baker. Maybe he could have made a living at that.

1 (9-inch) round loaf

2 cups water

½ cup yellow cornmeal

1 teaspoon salt

2 tablespoons butter

¼ cup light molasses

3½ to 4 cups bread or all-purpose flour

1 package (2¼ teaspoons) fast-rising yeast

1. Combine the water, cornmeal, and salt in a 1-quart saucepan. Bring the mixture to a boil over medium heat, stirring constantly. Cook, stirring, 1 minute. Remove from heat. Stir in the butter and molasses; set aside to cool to 120° to 125° F.

2. Combine 3½ cups bread flour and the yeast in a large bowl. Stir the cornmeal mixture into the flour mixture until a soft, smooth dough forms.

3. Turn the dough out onto a lightly floured board and knead for 3 minutes, adding as much of the remaining bread flour as necessary to make the dough manageable. Shape the dough into a ball on the board and cover it with the bowl. Let rise until double in bulk, 20 to 25 minutes.

4. When the dough has risen, grease a 9-inch round cake pan. Shape the dough into a ball and fit into the greased pan. Cover the loaf lightly with a linen towel and set aside in a warm place to rise until double in bulk, 30 to 35 minutes.

5. Preheat the oven to 375° F. Bake the loaf until it sounds hollow when gently tapped on the top, 30 to 35 minutes. Cool in the pan 5 minutes. Remove to a wire rack and cool to room temperature before slicing.

boston brown bread

Traditionally, Boston brown bread was made with equal parts cornmeal, rye flour, and Graham or whole wheat flour, but triticale, the modern hybrid of wheat and rye, makes an equally delicious loaf.

1 (1-pound) loaf

1. Generously grease a $1^1/2$ quart melon mold, pudding mold, or metal bowl. Place a rack in a pot large enough to hold the mold.

2. Combine the flour, cornmeal, sugar, baking soda, and salt in a medium bowl. Stir in the buttermilk, molasses, and raisins until well blended. Turn the mixture into the prepared mold. Cover the mold tightly with foil or its lid.

3. Place the mold on the rack in the pot. Fill the pot with water to reach halfway up the side of the mold. Bring the water to a boil over high heat. Reduce heat until the water

1½ cups triticale flour (see note)

1 cup cornmeal

2 tablespoons sugar

1½ teaspoons baking soda

½ teaspoon salt

1¼ cups buttermilk

¼ cup light molasses

¼ cup raisins

simmers. Cover the pot and steam the bread $1^{1}/_{2}$ to 2 hours or until a cake tester inserted into the center comes out clean, adding water as needed.

4. Cool for 10 minutes in the mold on a wire rack, then turn out and allow surface to air dry for 20 minutes before slicing.

note: *If triticale flour is not available, use $^{3}/_{4}$ cup bread or all-purpose flour and $^{3}/_{4}$ cup rye flour.*

ezekiel flat bread

In Ezekiel 4:9 an angel instructs Ezekiel to "take thou also unto thee wheat, and barley, and beans, and lentils, and millet, and fitches, and put them in one vessel, and make thee bread thereof." The resulting bread is dense, tender, and flavorful.

1 (14-inch) round loaf

3 to 3½ cups Ezekiel or multi-blend flour (see note)

2 tablespoons sugar

1 package (2¼ teaspoons) fast-rising yeast

¾ teaspoon salt

1 cup hot water (120° to 125° F)

2 tablespoons olive oil

¼ teaspoon coarse (kosher) salt (optional)

1. Combine 3 cups flour, the sugar, yeast, and salt in a large bowl. Stir the water and oil into the flour mixture until a soft smooth dough forms.

2. Turn the dough out onto a lightly floured board and knead for 3 minutes, adding as much of the remaining flour as necessary to make the dough manageable. Shape the dough into a ball on the board and cover it with the bowl. Let rise until double in bulk, 20 to 25 minutes.

3. When the dough has risen, grease a 14-inch round in the center of a baking sheet. Roll out the dough to a 14-inch round on the greased baking sheet. Cover the dough lightly with a linen towel and set aside in a warm place to rise until double in bulk, 25 to 30 minutes.

4. Preheat the oven to 375° F. Brush the surface of the dough with water, and if desired, sprinkle with kosher salt. Bake the loaf until it sounds hollow when gently tapped on the top, 20 to 25 minutes. Cool completely on a wire rack before slicing.

note: *Ezekiel flour may not have that name on the package. Look for the ingredients and the bible passage.*

lemon oatmeal bread

1 (9-inch) loaf

1. Preheat the oven to 350° F. Generously grease a 9 × 5-inch loaf pan.
2. Stir the flour, oats, baking powder, and salt in a medium bowl.
3. Beat the butter and sugar in a large bowl with an electric mixer. Beat in the eggs, one at a time, until a smooth batter forms.
4. At low speed, beat in the flour mixture alternately with the milk. Fold in the walnuts and lemon zest. Turn the mixture into the greased loaf pan.
5. Bake bread 55 to 60 minutes or until a cake tester inserted into the center comes out clean. Remove to a wire rack.
6. Meanwhile, prepare the Lemon Syrup. Combine the water, sugar, and lemon juice in a 1-quart saucepan. Bring to a boil.
7. When bread has been removed from oven, pierce it several times with a cake tester. Pour the syrup over the bread in the pan. Allow to stand 10 minutes, then remove from pan and cool completely before cutting.

Bread

1½ cups all-purpose flour

1 cup old-fashioned rolled oats

2 teaspoons baking powder

¼ teaspoon salt

8 tablespoons (1 stick) butter, softened

¼ cup sugar

2 eggs

½ cup milk

½ cup chopped walnuts

2 teaspoons grated lemon zest

Lemon Syrup

¼ cup water

3 tablespoons sugar

3 tablespoons lemon juice

blue-and-gold cornbread

A checkerboard bread, sure to be a conversation piece.

8 servings

1. Preheat the oven to 375° F. Grease a 9-inch square pan. Combine $1/2$ cup flour, the blue cornmeal, $1/2$ teaspoon salt, and 2 teaspoons baking powder in a bowl. Combine $1/2$ cup flour, the yellow cornmeal, $1/2$ teaspoon salt, and 2 teaspoons baking powder in another bowl.
2. Stir together the butter, buttermilk, egg, and maple syrup in a small bowl. Mix in the corn. Stir half into each of the cornmeal mixtures; mix well.
3. Drop tablespoonsful of alternate batters in a row until the pan is full. For instance, drop yellow, then blue, then yellow, and so on. Bake 20 minutes or until lightly browned.

1 cup all-purpose flour

$1/2$ cup blue cornmeal

$1/2$ cup yellow cornmeal

1 teaspoon salt

4 teaspoons baking powder

4 tablespoons butter, melted

1 cup buttermilk

1 egg

$1/4$ cup maple syrup or honey

2 cups corn kernels

rye and indian bread

During the Revolutionary War, German mercenary soldiers introduced the Hessian fly to New England wheat fields, making it difficult to grow wheat in the area for many years. This hearty bread became popular during that time. It uses rye flour and cornmeal (Indian meal) to extend the more expensive wheat flour, which was imported from the south.

1 (10-inch) loaf

1 cup rye flour

1 cup cornmeal

1 to 1½ cups bread flour

1 package (2¼ teaspoons) fast-rising yeast

¾ teaspoon salt

1 cup hot water (120° to 125° F.)

¼ cup light molasses

2 tablespoons vegetable oil

1. Combine the rye flour, cornmeal, 1 cup bread flour, the yeast, and salt in a large bowl. Stir the water, molasses, and oil into the flour mixture until a soft smooth dough forms.

2. Turn the dough onto a lightly floured board and knead for 3 minutes, adding as much of the remaining bread flour as necessary to make the dough manageable. Shape the dough into a ball on the board and cover it with the bowl. Let rise until double in bulk, 20 to 25 minutes.

3. When the dough has risen, grease a 10 × 6-inch oval in the center of a baking sheet. Shape the dough into a 9 × 5-inch oval and place it on the greased baking sheet. Cover the loaf lightly with a linen towel and set aside in a warm place to rise until double in bulk, 35 to 40 minutes.

4. Preheat the oven to 375° F. Bake the loaf until it sounds hollow when gently tapped on the top, 30 to 35 minutes. Cool completely on a wire rack before slicing.

whole wheat pretzels

Basically a form of bread, pretzels are sold the world over as a snack and in America as street food. Buy one on the streets of Philadelphia, Boston, or New York. Or better yet, make your own and turn the process into a family project—it's lots of fun.

12 pretzels

1. Combine 2 cups all-purpose flour, the whole wheat flour, yeast, salt, and sugar in a large bowl. Add the 1 cup hot water and mix until a soft dough forms.

2. Turn the dough out onto a lightly floured board and knead for 5 minutes, adding as much of the remaining $1/2$ cup all-purpose flour as necessary to make the dough manageable. Shape the dough into a ball on the board, cover it with the bowl, and let rise until double in bulk, 20 to 25 minutes.

3. Divide the dough into 12 equal pieces; roll each piece into a 12-inch

2 to 2½ cups all-purpose flour

1 cup whole wheat flour

1 package (2¼ teaspoons) fast-rising yeast

1½ teaspoons salt

1 teaspoon sugar

1 cup hot water (about 120° to 125° F.)

4 cups water

4 teaspoons baking soda

2 teaspoons cream of tartar

1 egg beaten with 1 tablespoon water

Coarse (kosher) salt

Wheat berries

rope. To form pretzels, place dough rope on a flat work surface in a **U** shape, like a horseshoe, with the ends toward you. Lift the ends, twist, and turn them back toward the bend of the **U**. Spread the ends apart and press onto the dough about $^1/_4$-inch on either side of the center of the bend. Set aside.

4. Preheat the oven to 475° F. Grease baking sheet.

5. Bring 4 cups water, the baking soda, and cream of tartar to a boil in a nonaluminum saucepan. Carefully drop 2 to 3 pretzels at a time into the water until they rise, 2 to 3 minutes. Remove with a slotted spoon and place on the prepared baking sheet. Brush with the egg and water mixture and sprinkle with salt and/or wheat berries.

6. Bake until golden brown, about 12 minutes. Let cool on a rack.

semolina soda bread

This cross-cultural combination yields a delicious golden loaf.

1 (8-inch) round loaf

1. Preheat the oven to 375° F. Grease an 8-inch round in the center of a baking sheet.

2. Combine the flour, semolina, sugar, baking powder, baking soda, and salt in a large bowl. With a pastry blender or two knives, cut the butter into the flour mixture until coarse crumbs form. Stir in the currants.

3. With a fork, stir the buttermilk into the flour mixture until a soft dough forms. Turn the dough out onto a lightly floured board and knead several times. Shape into a ball and place on a greased baking sheet. With a sharp knife cut an *X* into the top of the bread.

4. Bake the soda bread 30 to 40 minutes or until it sounds hollow when the top is tapped. Remove to a wire rack to cool to room temperature before cutting.

1 1/4 cups all-purpose flour

3/4 cup semolina

3 tablespoons sugar

3 teaspoons baking powder

1/2 teaspoon baking soda

1/2 teaspoon salt

6 tablespoons (3/4 stick) butter

2 tablespoons dried currants

1 cup buttermilk

desserts

Gin a body meet a

body

Comin thro' the rye,

Gin a body kiss a

body

Need a body cry?

—ROBERT BURNS,

Comin Thro' the Rye,

1796

apple-cranberry crisp

Crisps are one of the traditional ways to use rolled oats in baking.

6 servings

1. Combine the apple juice, 1/4 cup brown sugar, the lemon juice, cornstarch, lemon zest, and nutmeg in a 2-quart saucepan. Bring to a boil over medium-high heat, stirring constantly, until thickened, about 4 minutes. Fold in the apples and cranberries. Set aside.

2. Combine the flour, oats, remaining 1/4 cup brown sugar, and the salt in a medium bowl. Stir in the melted butter until the mixture forms coarse crumbs.

3. Preheat the oven to 350° F. Grease an 8-inch square baking pan. Pat half of the oatmeal mixture into the baking pan to cover the bottom. Spoon the apple-cranberry mixture over the crumbs, smoothing to make a level surface. Top with the remaining crumb mixture and the walnuts.

4. Bake until the filling is bubbly and the topping is golden brown, 35 to 40 minutes. Cool 15 minutes, then cut into 6 rectangles. Serve warm or at room temperature, with ice cream or whipped cream if desired.

1 cup apple juice or cider

1/2 cup firmly packed light brown sugar

3 tablespoons freshly squeezed lemon juice

1 tablespoon cornstarch

1 teaspoon grated lemon zest

1/4 teaspoon freshly grated nutmeg

3 cups peeled and thinly sliced apples (about 2 large apples)

1 cup fresh or thawed frozen cranberries

1 cup all-purpose flour

1 cup old-fashioned rolled oats

1/4 teaspoon salt

6 tablespoons (3/4 stick) butter, melted

1/4 cup chopped walnuts or pecans

Vanilla ice cream or whipped cream (optional)

applesauce oatmeal cake

A simple snack cake the whole family is sure to love.

1 (8-inch) square cake

1. Preheat the oven to 350° F. Grease an 8-inch square pan.

2. Combine the flour, oats, sugar, baking powder, baking soda, zests, cinnamon, ginger, nutmeg, and salt in a medium bowl.

3. Whisk together the eggs and oil in a small bowl, then add the lemon and orange juices. Add to the dry ingredients. Mix in the applesauce and almonds. Pour into the prepared pan. Bake 30 minutes or until a cake tester inserted in the center comes out clean. Cool on a rack.

variation: *For chocolate chip applesauce snack cake: add 1 cup mini-chocolate chips to the dry ingredients. Or remove the cake from the oven, sprinkle the chips on top, return to the oven for 1 minute, remove, and spread the melted chocolate.*

1 cup all-purpose flour

½ cup old-fashioned rolled oats

¾ cup sugar

1 teaspoon baking powder

1 teaspoon baking soda

1 teaspoon grated orange zest

1 teaspoon grated lemon zest

½ teaspoon ground cinnamon

¼ teaspoon ground ginger

⅛ teaspoon freshly grated nutmeg

¼ teaspoon salt

2 eggs

3 tablespoons vegetable oil

2 tablespoons freshly squeezed lemon juice

2 tablespoons orange juice

½ cup unsweetened applesauce

½ cup chopped almonds

granola chocolate chippers

Full of grain goodness, these hearty and tasty bars will please any palate.

25 squares

1. Preheat the oven to 350° F. Grease a 9-inch square pan. Beat butter and sugar in a medium bowl with an electric mixer on medium speed until light and fluffy. Add the egg and vanilla.

2. Stir in Granola, flours, baking powder, and salt until blended. Mix in chocolate chips. Turn into the prepared pan.

3. Bake 18 to 20 minutes or until a toothpick inserted in the center comes out with a few crumbs clinging to it. Remove the pan to a rack to cool completely before cutting into 25 squares.

8 tablespoons (1 stick) butter, softened

¾ cup firmly packed dark brown sugar

1 egg, room temperature

1 teaspoon vanilla extract

1 cup Granola (see page 25)

¾ cup all-purpose flour

¼ cup whole wheat flour

1 teaspoon baking powder

½ teaspoon salt

1 (6-ounce) bag chocolate chips

indian pudding
with apples

Although not native to America, apples make a delicious addition to this traditional dessert.

6 servings

1. Preheat the oven to 325° F. Generously grease a shallow 1¹/₂-quart baking dish. Peel, core, and coarsely chop the apples.

2. Bring the milk and cornmeal to a boil in a 2-quart saucepan over medium heat, stirring constantly. Cook, stirring, 5 minutes. Stir in the apples, molasses, ginger, cinnamon, and salt. Turn the mixture into the prepared dish. Pour the cream over the top of the cornmeal mixture and sprinkle with the brown sugar.

3. Place on a rimmed baking sheet. Bake for 1 to 1¹/₂ hours or until cream has been absorbed and pudding is firm. Serve warm.

2 medium apples

3 cups milk

½ cup yellow cornmeal

¼ cup molasses

1 teaspoon ground ginger

2 teaspoons ground cinnamon

½ teaspoon salt

½ cup heavy cream

2 tablespoons packed light brown sugar

irish oats pudding

6 servings

½ cup steel-cut (Irish) oats

1 cup boiling water

1 egg

1 cup milk

1 cup heavy cream

¼ cup sugar

2 teaspoons vanilla extract

¼ teaspoon salt

¼ cup dried sour cherries or raisins

Ground cinnamon

1. Toast the oats in a large dry skillet over medium heat, stirring constantly for 5 minutes. Add the water and return to a boil over high heat; remove skillet from heat, cover, and set aside 20 minutes or until the oats have absorbed all of the water.

2. Preheat the oven to 350° F. Lightly grease a 1½-quart shallow casserole. Beat the egg until frothy in a medium bowl. Beat in the milk, cream, sugar, vanilla, and salt. Stir in the oats and cherries.

3. Turn the oat mixture into the prepared casserole and sprinkle the cinnamon over the top.

4. Bake 45 to 50 minutes or until set and lightly browned. Allow pudding to cool 15 to 20 minutes at room temperature, then serve.

mixed-grain cocoa cake

9 servings

1. Preheat the oven to 350° F. Lightly grease a 9-inch square baking pan.

2. Combine the flour, sugar, and salt in a large bowl. Cut the butter into the flour mixture with a pastry blender or two knives until the mixture is crumbly. Stir in the wheat, oats, and rye. Remove ¼ cup of the mixture. Stir the cocoa, baking powder, and baking soda into the remaining mixture in the bowl.

3. Combine the milk and vanilla in a small bowl; then stir into the flour mixture until well blended. Turn the mixture into the prepared pan. Sprinkle the reserved crumbs over top.

4. Bake for 35 to 40 minutes or until crumbs are golden brown and center is firm when gently pressed.

5. Cool for 20 minutes in the pan on a wire rack, then cut into 9 pieces and serve warm.

1 cup all-purpose flour

1 cup sugar

½ teaspoon salt

8 tablespoons (1 stick) butter, sliced

¼ cup rolled wheat

¼ cup rolled oats

¼ cup rolled rye, triticale, or kamut

⅓ cup unsweetened cocoa

2 teaspoons baking powder

¼ teaspoon baking soda

1 cup milk

1½ teaspoons vanilla extract

quick couscous pudding

6 servings

1. Combine the couscous, hot milk, sugar, and raisins in a medium bowl. Set aside 20 minutes or until the couscous absorbs all of the milk.
2. Stir in 8-ounces yogurt. Sprinkle the top of the pudding with cinnamon; serve warm or cover and refrigerate until ready to serve. If pudding is not served within 2 or 3 hours, it may be necessary to stir in additional yogurt before serving.

½ cup couscous

1 cup skim milk, heated just to boiling

2 tablespoons sugar

¼ cup raisins

1 (8-ounce) container of vanilla-flavored low-fat yogurt

Ground cinnamon

Additional vanilla-flavored low-fat yogurt (optional)

mixed-grain peanut butter cookies

About 3 dozen

1. Preheat the oven to 350° F. Lightly grease 2 baking sheets.

2. Beat the sugar, butter, and peanut butter in a large bowl with an electric mixer set on medium speed until well blended. Beat in the egg and vanilla until light and fluffy.

3. Beat in the flour, baking powder, and salt, until the dough is smooth, scraping the sides of the bowl frequently with a rubber scraper. Stir in the oats, wheat, and rye.

4. Drop the dough by heaping teaspoonfuls 2 inches apart onto the baking sheets. Press the dough with the floured back of a spoon to flatten slightly. Bake 12 to 15 minutes or until edges are golden brown. Cool on a wire rack. Store cookies in a tight container in a cool, dry place.

note: *If you cannot find the recommended rolled or flaked grain products, make your own mixture of rolled grains, use all rolled oats, or use rolled mixed-grain hot cereal.*

¾ cup firmly packed light brown sugar

4 tablespoons (½ stick) butter, softened

½ cup chunky peanut butter

1 egg

1 teaspoon vanilla extract

¾ cup all-purpose flour

1 teaspoon baking powder

¼ teaspoon salt

⅓ cup rolled oats (see note)

⅓ cup wheat flakes (see note)

⅓ cup rye or barley flakes (see note)

soft oatmeal
sugar cookies

About 20

1. Preheat the oven to 350° F.
Lightly grease 2 baking sheets.
2. Beat the granulated sugar, butter, egg, and vanilla in a large bowl with an electric mixer set on medium speed until well blended. Beat in the milk, then the flour, baking powder, cinnamon, and salt until a stiff batter forms. Fold in the oats and raisins.
3. Drop by heaping measuring tablespoonfuls 3 inches apart onto the prepared baking sheets. Bake 18 to 20 minutes or until the edges are golden brown. Let cookies cool 2 minutes on sheets, then remove to racks to cool completely. Cookies are best if eaten within a day or so of baking. Store in a tight container in a cool, dry place. Dust with confectioners' sugar before serving.

1 cup granulated sugar

1/2 cup melted butter
or vegetable
shortening

1 egg

2 teaspoons vanilla
extract

1/2 cup milk

1 1/4 cups all-purpose
flour

1 teaspoon baking
powder

1/2 teaspoon ground
cinnamon

1/4 teaspoon salt

1 cup old-fashioned
rolled oats

1/4 cup raisins

Confectioners' sugar

whole-grain molasses cookies

About 3 dozen

1. Preheat the oven to 350° F. Lightly grease 2 baking sheets.
2. Stir together the flour, oats, sugar, dates, cinnamon, ginger, baking soda, and salt in a medium bowl.
3. Combine the oil, molasses, and egg in a glass measuring cup. Make a well in the center of the flour mixture. Pour the molasses mixture into the well and stir with a fork until well blended.
4. Drop the dough by teaspoonfuls onto the prepared baking sheets and bake 10 to 12 minutes or until the edges start to brown and the centers feel firm. Cool on a wire rack. Store cookies in a tight container in a cool, dry place.

1 cup whole wheat flour

1 cup old-fashioned rolled oats

¼ cup firmly packed light brown sugar

¼ cup chopped dates

½ teaspoon ground cinnamon

½ teaspoon ground ginger

¼ teaspoon baking soda

¼ teaspoon salt

⅓ cup vegetable oil

¼ cup molasses

1 egg

mail order sources

If you can look into

the seeds of time,

And say which grain

will grow and which

will not,

Speak then to me,

who neither beg nor

fear

Your favors nor your

hate.

—WILLIAM

SHAKESPEARE,

Macbeth, 1606

Arrowhead Mills
Box 2059
Hereford, TX 79045-2059
(806) 364-0730

Balducci's
424 Avenue of the Americas
New York, NY 10011
(800) 225-3822

Dean & Deluca
560 Broadway
New York, NY 10012
(212) 431-1691
(800) 221-7714 (outside NY)

Kalustyan's
123 Lexington Avenue
New York, NY 10016
(212) 685-3451
(212) 685-3888
(212) 683-8458 (fax)

King Arthur Flour Co.
Rural Route 2, Box 56
Norwich, VT 05055
(800) 827-6836

Shiloh Farms
P.O. Box 97
Benton County
Sulphur Springs,
AR 72768-0097
(800) 362-6832

Walnut Acres
Penns Creek, PA 17862
(800) 433-3998

grain finder

Mares eat oats

And does eat oats

And little lambs eat

ivy.

—MILTON DRAKE,

Mairzy Doats, 1943

A quick glance at the following list will tell you which recipes you can make with the grain you want to use.

amaranth

Fresh corn and amaranth
  tamales
Mixed-grain and wild mushroom pilaf
Mixed-grain pilaf with fruit

barley

Barley and bell pepper cakes
Barley and tricolored pepper
  salad
Barley Provençale
Barley with fennel
Beef and barley soup
Ezekiel flat bread
Granola
Granola chocolate chippers
Hungarian goulash soup with
  barley and rice
Mixed-grain and vegetable pie
Mixed-grain peanut butter
  cookies
Mushroom barley pilaf
Smoked turkey, herbed barley,
  and vegetable salad
Squash and barley casserole
White bean and barley soup
Wild mushroom and barley
  soup

buckwheat

Buckwheat blini

Buckwheat noodles in spicy
  peanut sauce
Buckwheat waffles
Green kasha pilaf
Kasha varnishkes
Uncle Herb's brisket with kasha
  and rice

corn

Anadama bread
Black bean, red pepper, corn,
  and quinoa salad
Blue-and-gold cornbread
Blueberry corn muffins
Boston brown bread
Chicken posole
Corn dodgers
Corn, millet, and crab chowder
Corn, quinoa, and red pepper
  chili chowder
Fresh corn and amaranth
  tamales
Fresh corn empanadas
Grits and grains soufflé
Hominy and sausage
Indian pudding with apples
Lacy cornmeal pancakes
Red pepper polenta with
  ratatouille
Rye and Indian bread
Scrapple
Three-grain hush puppies

kamut

Kamut tagliatelle with roasted
   vegetables
Mixed-grain cocoa cake (option)

millet

Corn, millet, and crab chowder
Ezekiel flat bread
Mixed-grain pilaf with fruit
Mixed-grain and vegetable pie
Mixed-grain and wild mush-
   room pilaf

oats

Apple-cranberry crisp
Applesauce oatmeal cake
Buttermilk oatmeal pancakes
Granola
Granola chocolate chippers
Irish oatmeal pilaf
Irish oats pudding
Irish ris-oat-o
Lamb and oats stew
Lemon oatmeal bread
Mixed-grain and vegetable
   pie
Mixed-grain cocoa cake
Mixed-grain peanut butter
   cookies
Muesli
Soft oatmeal sugar cookies
Three-grain hush puppies
Whole-grain molasses cookies

quinoa

Aromatic chicken and quinoa
Black bean, red pepper, corn,
and quinoa salad
Corn, quinoa, and red pepper
   chili chowder
Hummus with quinoa
Quinoa and wild and brown
   rice pilaf

rice

Hungarian goulash soup with
   barley and rice
Mixed-grain and wild mushroom pilaf
Mixed-grain pilaf with fruit
Quinoa and wild and brown
   rice pilaf
Uncle Herb's brisket with kasha
   and rice

rye

Corn dodgers
Granola
Granola chocolate chippers
Mixed-grain cocoa cake
Mixed-grain peanut butter
   cookies
Rye and Indian bread
Sweet-and-sour rye meatballs

spelt

Three-grain hush puppies

teff

Grits and grains soufflé
Mixed-grain pilaf with fruit

triticale

Boston brown bread
Mixed-grain cocoa cake (option)

wheat

bulgur

Bulgur and chicken salad
Bulgur and vermicelli pilaf
Bulgur garden vegetable salad
Chicken with sun-dried tomatoes, rice, and bulgur
Gazpacho bulgur salad
Mixed-grain and wild mushroom pilaf
Tabbouleh
Wheat and turkey loaf

couscous

Athenian couscous salad
Citrus couscous salad
Couscous and chicken madeira
Moroccan lamb stew
Quick couscous pudding

flour

Anadama bread
Applesauce oatmeal cake
Barley and bell pepper cakes
Blue-and-gold cornbread
Blueberry corn muffins
Buckwheat blini
Buckwheat waffles
Buttermilk oatmeal pancakes
Corn dodgers
Fresh corn empanadas
Granola chocolate chippers

Lacy cornmeal pancakes
Lemon oatmeal bread
Mixed-grain and vegetable pie
Mixed-grain cocoa cake
Mixed-grain peanut butter cookies
Rye and Indian bread
Semolina soda bread
Soft oatmeal sugar cookies
Three-grain hush puppies
Whole wheat pretzels
Whole-grain molasses cookies

semolina

Bulgur and vermicelli pilaf (pasta)
Kasha varnishkes (pasta)
Parmesan gnocchi
Semolina soda bread

other wheat products

Granola (germ or bran)
Granola chocolate chippers (germ or bran)
Grits and grains soufflé (cereal)
Lentil and warm wheat berry salad (berries)
Mixed-grain cocoa cake (rolled)
Mixed-grain peanut butter cookies (rolled)
Wheat and turkey loaf (berries)

index

conversion chart
equivalent imperial and metric measurements

American cooks use standard containers, the 8-ounce cup and a tablespoon that takes exactly 16 level fillings to fill that cup level. Measuring by cup makes it very difficult to give weight equivalents, as a cup of densely packed butter will weigh considerably more than a cup of flour. The easiest way therefore to deal with cup measurements in recipes is to take the amount by volume rather than by weight. Thus the equation reads:

$$1 \text{ cup} = 240 \text{ ml} = 8 \text{ fl. oz.} \quad \tfrac{1}{2} \text{ cup} = 120 \text{ ml} = 4 \text{ fl. oz.}$$

It is possible to buy a set of American cup measures in major stores around the world.

In the States, butter is often measured in sticks. One stick is the equivalent of 8 tablespoons. One tablespoon of butter is therefore the equivalent to ½ ounce/15 grams.

liquid measures

Fluid ounces	U.S.	Imperial	Milliliters
	1 tsp	1 tsp	5
¼	2 tsp	1 dessertspoon	10
½	1 tbs	1 tbs	15
1	2 tbs	2 tbs	28
2	¼ cup	4 tbs	56
4	½ cup		110
5		¼ pint or 1 gill	140
6	¾ cup		170
8	1 cup		225
9			250, ¼ liter
10	1¼ cups	½ pint	280
12	1½ cups		340
15		¾ pint	420
16	2 cups		450
18	2¼ cups		500, ½ liter
20	2½ cups	1 pint	560
24	3 cups		675
25		1¼ pints	700
27	3½ cups		750
30	3¾ cups	1½ pints	840
32	4 cups		900
35		1¾ pints	980
36	4½ cups		1000, 1 liter
40	5 cups	2 pints	1120
48	6 cups		1350
50		2½ pints	1400
60	7½ cups	3 pints	1680
64	8 cups		1800
72	9 cups		2000, 2 liters

solid measures

U.S. and Imperial Measures		Metric Measures	
ounces	pounds	grams	kilos
1		28	
2		56	
3½		100	
4	¼	112	
5		140	
6		168	
8	½	225	
9		250	¼
12	¾	340	
16	1	450	
18		500	½
20	1¼	560	
24	1½	675	
27		750	¾
28	1¾	780	
32	2	900	
36	2¼	1000	1
40	2½	1100	
48	3	1350	
54		1500	1½
64	4	1800	
72	4½	2000	2
80	5	2250	2¼
90		2500	2½
100	6	2800	2¾

oven temperature equivalents

Fahrenheit	Celsius	Gas Mark	Description
225	110	¼	Cool
250	130	½	
275	140	1	Very Slow
300	150	2	
325	170	3	Slow
350	180	4	Moderate
375	190	5	
400	200	6	Moderately Hot
425	220	7	Fairly Hot
450	230	8	Hot
475	240	9	Very Hot
500	250	10	Extremely Hot

Any broiling recipes can be used with the grill of the oven, but beware of high-temperature grills.

suggested equivalents and substitutes for ingredients

all-purpose flour—plain flour
baking sheet—oven tray
buttermilk—ordinary milk
cheesecloth—muslin
coarse salt—kitchen salt
confectioner's sugar—icing sugar
cornstarch—cornflour
granulated sugar—caster sugar
half and half—12% fat milk

heavy cream—double cream
light cream—single cream
plastic wrap—cling film
shortening—white fat
unbleached flour—strong, white flour
vanilla bean—vanilla pod
zest—rind
zucchini—courgettes or marrow